PRAISE FOR

BECOMING EARTH

"[Ferris Jabr] vividly describes how much life has shaped, and has been shaped by, our planet. . . . He writes eloquently [and] beautifully about his encounters around the world with biologists, environmentalists, farmers and others involved in some way with the life of the planet." —*The Wall Street Journal*

"A convincing, mind-opening case that 'the history of life on Earth is the history of life remaking Earth.'" —*The Atlantic*

"This exploration of the Earth as a living entity is lyrical, smart, and will make you appreciate the home planet in countless new ways. It's really a fascinating book." —NPR *Science Friday*

"Infectiously poetic . . . exhilarating . . . The overall uplifting outlook is sure to reenergize even the most hardened climate pessimist." —*Science*

"Wide-ranging and thought-provoking . . . The best books manage to entertain, educate, astonish and even galvanize the reader . . . They expand awareness [and] serve as celebrations and warnings, challenges and pleas. . . . With *Becoming Earth*, Oregon-based journalist Ferris Jabr achieves all of these aims and more." —*The Guardian*

"Convincing . . . poetic . . . punctuated with thunderstorm-like downpours of ideas."

—*Sierra*

"[Jabr] explores all the ways life has shaped our physical world and, in doing so, inevitably revisits the question 'Is the Earth alive?'" —*Vox*

"Ferris Jabr has an uncanny ability to explore and explain some of the greatest mysteries of the universe, and his sentences are both luscious and limpid. . . . Jabr is an exceptional new science writer, and this urgent book is poised to influence larger conversations about climate change and the environment."

—The Whiting Foundation Creative Nonfiction Grant Jury

"An electrifying debut that earns its place alongside the best of today's essential popular science books, as well as acknowledged classics."

—*Red Canary Magazine*

"A real feast for readers of books about life on Earth and for those who appreciate literary work: Jabr is not only knowledgeable, but a master of lyrical prose."

—*3 Quarks Daily*

"A masterwork of journalism—exhaustively researched, wide-ranging, simultaneously intricate in detail and accessible to general readers . . . This book will revolutionize readers' concepts of the fundamental interdependency of life, air, and soil. With the curiosity of a reporter, the mind of a scientist, and the lyricism of a poet, Jabr explores the extraordinary tapestry of life."

—*Kirkus Reviews* (starred review)

"An enlightening examination of how living organisms have influenced their environments . . . an edifying and holistic view of life on earth."

—*Publishers Weekly*

"A beautifully written, exquisitely detailed, and finely researched examination of life and its symbiotic relationship with Earth."

—*Booklist* (starred review)

"A master class in science journalism by one of the top practitioners of the form." —Charleston *Post and Courier*

"A compelling account of interconnectedness . . . Jabr weaves a tapestry out of the complex relationships that life forms have not only with each other, but also with the stuff of Earth itself." —*New Scientist*

"Wonderful . . . Jabr tells this story with admirable thoroughness. . . . *Becoming Earth* is a fantastic book for becoming aware of the great beauty and power residing on this planet we live on and to which we give so little attention." —*Big Think*

"Tremendous . . . Jabr endeavors to tell the story of life on planet Earth from the very beginning." —*Atmos*

"*Becoming Earth* is a glorious paean to our living world, full of achingly beautiful passages, mind-bending conceptual twists, and wonderful characters. Ferris Jabr reveals how Earth not only gave rise to life, and now teems with it, but has also been profoundly, miraculously shaped by it." —Ed Yong, author of *An Immense World*

"This is the book I'd been waiting for. It tells my favorite kind of science story: one that seems at first counterintuitive, but then quickly becomes obviously true—a story so important and compelling that I am going to be recommending it for years. It is absolutely packed with delicious nuggets of insight. I fell into this book, and when I emerged, it had deeply changed the way I see what is, by far, the best planet." —Hank Green, *New York Times* bestselling author and science communicator

"*Becoming Earth* is an astonishing book, weaving together science, history, and the author's unfailingly precise observations with the grace of a poet. To read it is to be transformed by a new understanding of our place in what Jabr calls the 'collaborative and improvised performance' that sustains our marvelous home in the universe."

—Steve Silberman, author of *NeuroTribes: The Legacy of Autism and the Future of Neurodiversity*

"As a writer, Jabr is chimerical: he has the sharp eyes of a scientist, the big ears of a journalist, and the liquid-silver tongue of a poet. There are times, reading this paradigm-shifting book, when you will feel like you are peering right down into the very heart of our living planet. It is, quite simply, a work of genius."

—Robert Moor, bestselling author of *On Trails: An Exploration*

"Ferris Jabr explores the many ways life has transformed the planet over the last three billion years. In the process, he offers a fresh perspective on today's most urgent challenges. *Becoming Earth* is fascinating, thought-provoking, and, ultimately, inspiring."

—Elizabeth Kolbert, Pulitzer Prize–winning author of *The Sixth Extinction*

"Gorgeously written and brimming with fascinating science and provocative ideas, *Becoming Earth* is a book about transformation that is itself transformative. Dive into it and you will come to see our world as Ferris Jabr sees it: a thrumming, breathing entity that doesn't merely harbor life, it *is* life."

—Dan Fagin, author of the Pulitzer Prize–winning *Toms River*

"We tend to take our rare jewel of a home planet for granted. In his startlingly beautiful and insightful book, *Becoming Earth,* Ferris Jabr shows us exactly why we shouldn't. The Earth lives, breathes, and rewrites our history even as we read, reminding us once again that there is in fact no place like home."

—Deborah Blum, Pulitzer Prize winner and author of *The Poison Squad*

"Wow. This wondrous book reveals our living planet for the miracle that it is. By the end, you may even feel that 'miracle' is an understatement. The story of Earth is the story of a planet reworked and remade—and, to an astonishing degree, created—by life itself."

—Carl Safina, author of *Beyond Words: What Animals Think and Feel*

"Thrilling . . . Ferris Jabr reveals the fascinating and cutting-edge science of life's intricate connections with Earth's chemistry, water cycle, rock record, soil, and air. And he shows us how these connections are profoundly relevant to human civilization. The result is more than excellent science journalism—it evokes wonder."

—Tyler Volk, Earth system scientist, New York University, and author of *Gaia's Body: Toward a Physiology of Earth*

"Ferris Jabr presents a fresh and compelling argument that Earth is a living planet. In the process, he takes you from the bottom of the mile-deep Homestake Mine in South Dakota to the top of the 1,066-foot Amazon Tall Tower Observatory in Brazil. *Becoming Earth* is a fascinating read that should help everyone appreciate the splendor of our natural environment."

—James Kasting, professor of geosciences, Penn State University

BECOMING EARTH

BECOMING EARTH

A Journey Through the Hidden Wonders That Bring Our Planet to Life

FERRIS JABR

RANDOM HOUSE
NEW YORK

2025 Random House Trade Paperback Edition

Published in the United States by Random House, an imprint and division of Penguin Random House LLC, 1745 Broadway, New York, NY 10019.

Random House and the House colophon are registered trademarks of Penguin Random House LLC.

Random House Book Club and colophon are trademarks of Penguin Random House LLC.

Originally published in hardcover in the United States by Random House, an imprint and division of Penguin Random House LLC, in 2024.

Library of Congress Cataloging-in-Publication Data

Names: Jabr, Ferris, author.
Title: Becoming Earth: a journey through the hidden wonders that bring our planet to life / Ferris Jabr.
Description: New York: Random House, [2023] |
Includes bibliographical references and index.
Identifiers: LCCN 2023038723 (print) | LCCN 2023038724 (ebook) |
ISBN 9780593133996 (trade paperback) | ISBN 9780593133989 (ebook)
Subjects: LCSH: Life (Biology) | Evolution (Biology) | Life—Origin.
Classification: LCC QH367 .J23 2023 (print) | LCC QH367 (ebook) |
DDC 570.1—dc23/eng/20240201
LC record available at lccn.loc.gov/2023038723
LC ebook record available at lccn.loc.gov/2023038724

Printed in the United States of America on acid-free paper

randomhousebooks.com
randomhousebookclub.com
penguinrandomhouse.com

2 4 6 8 9 7 5 3 1

Book design by Barbara M. Bachman

The authorized representative in the EU for product safety and compliance is Penguin Random House Ireland, Morrison Chambers, 32 Nassau Street, Dublin D02 YH68, Ireland, https://eu-contact.penguin.ie.

FOR AIR, WATER, AND ROCK. FOR FIRE, ICE, AND MUD.

For the gnawing glaciers, rippling dunes, prismatic hot springs, and abyssal plains.

For the searing undersea vents, explosive magma chambers, ancient mountains, and newborn islands. For the great green forests, billowing grasslands, and spongy peat;

the abrupt plateaus, treeless tundra, and salt-soaked mangroves.

FOR THE DINOSAURS, REDWOODS, MAMMOTHS, AND WHALES;

the slime molds, insects, fungi, and snails.

For the microbes that eat sunlight, seed clouds, and mine gold.

For the roots that gave us soil and made rivers flow.

For the herds of extinct titans and all those that still roam.

For the ocean in our blood and our skeletons of stone.

FOR THE GROWERS, BUILDERS, THINKERS, AND TEACHERS.

For the explorers, creators, carers, and healers.

For all the songs we know and all those still unheard.

For our living planet. For our miracle. For Earth.

Consider: Whether you're a human being, an insect, a microbe, or a stone, this verse is true. All that you touch, You Change. All that you Change, Changes you.

—Octavia Butler, *The Parable of the Sower*

What if we understood that the heartbeats of all creatures can be heard in our own heartbeats, that this drumming is an echo of the Earth's pulse, which runs through all of our veins, plants and animals included?

—Terry Tempest Williams,
"Take Place," *The Paris Review*

Earth is one country. We are all waves of the same sea, leaves of the same tree, flowers of the same garden.

—The Message of Fraternity, likely
paraphrased from the writings of
Bahá'u'lláh, prophet-founder of the
Bahá'í Faith, and his son, 'Abdu'l-Bahá

CONTENTS

AIR

INTRODUCTION

WHEN I WAS A BOY, I THOUGHT I COULD CHANGE THE WEATHER. On sweltering summer days in suburban California, when gardens wilted and asphalt burned bare skin, I would draw a picture of a stout blue rain cloud and march around it on the lawn, splashing it with a potion of hose water and yard trimmings. I may have even sung a rudimentary incantation—an inversion of the popular nursery rhyme that tells rain to "go away."

As I grew up, so did my understanding of meteorology. In school, I learned how water evaporates from lakes, rivers, and oceans and rises into the atmosphere, where it cools and condenses into minuscule droplets. These drifting beads of water collide and coalesce, clinging to bits of floating dust and growing into the cottony masses we call clouds, which eventually become heavy enough to fall back to the surface. Rain, I was taught, is an inevitable outcome of atmospheric physics—a gift that we and other living creatures passively receive.

Some years ago, however, I learned a startling fact that completely changed the way I think about weather and ultimately altered my perception of the planet as a whole—a fact that returned me to a sense of awe and possibility I had rarely felt since childhood. What I learned is this: in many cases, life does not simply receive rain—it summons it.

Consider the Amazon rainforest. Every year, the Amazon is drenched in about eight feet of rain. In some parts of the forest, the annual rainfall is closer to fourteen feet—more than five times the average yearly precipitation across the contiguous United States. This deluge is partly a consequence of geographic serendipity: intense equatorial sunlight speeds the evaporation of water from sea and land to sky, trade winds bring moisture from the ocean, and bordering mountains force incoming air to rise, cool, and condense. Rainforests happen where it happens to rain.

But that's only half of the story. Within the forest floor, vast symbiotic networks of plant roots and filamentous fungi pull water from the soil into trunks, stems, and leaves. As the nearly 400 billion trees in the Amazon drink their fill, they release excess moisture, saturating the air with 20 billion tons of water vapor each day. At the same time, plants of all kinds secrete salts and emit bouquets of pungent gaseous compounds. Mushrooms, dainty as paper parasols or squat as doorknobs, exhale plumes of spores. The wind sweeps bacteria, pollen grains, and bits of leaves and bark into the atmosphere. The wet breath of the forest—peppered with microscopic life and organic residues—creates the ideal conditions for rain. With so much water in the air and so many minute particles on which the water can condense, clouds quickly form. Certain airborne bacteria even encourage water droplets to freeze, making clouds bigger, heavier, and more likely to burst. In a typical year, the Amazon generates around half of its own rainfall.

Ultimately, the Amazon rainforest influences much more than the weather above its canopy. All the water, biological detritus, and microscopic beings discharged by the forest form a massive floating river—an aerial echo of the one snaking through the understory. This flying river brings precipitation to farms and cities throughout South America. Some scientists have concluded that, due to long-range atmospheric ripple effects, the Amazon contributes to rainfall in places as far away as Canada. A tree growing in Brazil can change the weather in Manitoba.

The Amazon's secret rain ritual challenges the way we typically think about life on Earth. Conventional wisdom holds that life is subject to its environment. If Earth did not orbit a star of the right size and age, if it were too close or too far from that star—if it did not have a stable atmosphere, liquid water, and a magnetic field that deflects harmful cosmic rays—it would be lifeless. Life evolved on Earth because Earth is suitable for life. Since Darwin, prevailing scientific paradigms have likewise emphasized that the ever-shifting demands of the environment largely dictate how life evolves: species best able to cope with changes to their particular habitats leave behind the most descendants, whereas those that fail to adapt die out.

Yet this truth has an underappreciated twin: life changes its environment, too. In the mid–twentieth century, when ecology established itself as a formal discipline, this fact began to gain wider recognition in Western science. Even so, the focus was on relatively small and local changes: a beaver constructing a dam, for instance, or earthworms churning a patch of soil. The notion that living creatures of all kinds might modify their environments in much more dramatic ways—that microbes, fungi, plants, and animals can change the topography and climate of a continent or even the entire planet—was rarely given serious consideration. "To a large extent, the physical form and the habits of the earth's vegetation and its animal life have been molded by the environment," Rachel Carson wrote in *Silent Spring* in 1962. "Considering the whole span of earthly time, the opposite effect, in which life actually modified its surroundings, has been relatively slight." E. O. Wilson declared something similar in his 2002 book, *The Future of Life:* "*Homo sapiens* has become a geophysical force, the first species in the history of the planet to attain that dubious distinction."

When I first learned about the Amazon's rain dance, I was both enthralled and bewildered. I knew that plants pulled water from the ground and expelled moisture into the air, but the fact that trees, fungi, and microbes in the Amazon collectively conjured so much of the rain for which their home was named, that life's activity on one continent altered weather on another, was jolting. The idea of the Amazon rainforest as a garden that watered itself obsessed me. If that were true for an ecosystem as massive as the Amazon, I wondered, might it be true on an even larger scale? In what ways, and to what extent, has life changed the planet throughout its history?

In seeking the answers to these questions, I've learned that the scientific understanding of life's relationship to the planet has been undergoing a major reformation for some time now. Contrary to longstanding maxims, life has been a formidable geological force throughout Earth's history, often matching or surpassing the power of glaciers, earthquakes, and volcanoes. Over the past several billion years, all manner of life forms, from microbes to mammoths, have transformed the continents, ocean, and atmosphere, turning a lump of orbiting rock into the world as we've known it. Living creatures are not simply products of inexorable evolutionary processes in their particular habitats; they are orchestrators of their environments and participants in their own evolution. We and other living creatures are more than inhabitants of Earth; we *are* Earth—an outgrowth of its physical structure and an engine of its global cycles. Earth and its creatures are so closely intertwined that we can think of them as one.

The evidence for this new paradigm is all around us, although much of it has been discovered only recently and has yet to permeate public consciousness to the same degree as, say, selfish genes or the microbiome. Nearly two and a half billion years ago, photosynthetic ocean microbes called cyanobacteria permanently altered the planet, suffusing the atmosphere with oxygen, imbuing the sky with its familiar blue hue, and initiating the formation of the ozone layer, which protected new waves of life from harmful exposure to ultraviolet radiation. Today, plants and other photosynthetic organisms help maintain a level of atmospheric oxygen high enough to support complex life but not so high that that Earth would erupt in flames at the slightest spark. Microorganisms are important participants in many geological processes, responsible for much of Earth's mineral diversity; some scientists think they played a crucial role in forming the continents. Marine plankton drive chemical cycles on which all other life depends and emit gases that increase cloud cover, modifying global climate. Kelp forests, coral reefs, and shellfish store huge amounts of carbon, buffer ocean acidity, improve water quality, and defend shorelines from severe weather. And animals as diverse as elephants, prairie dogs, and termites continually reconstruct the planet's crust, facilitating the flow of water, air, and nutrients and improving the prospects of millions of species.

Humans are the most extreme example of life transforming Earth in recent history—and, by some measures, the most extreme ever. By exhuming and burning the carbon-rich remains of ancient jungles and sea creatures in the form of fossil fuels, as well as by destroying and degrading ecosystems, industrialized nations have flooded the atmosphere with carbon dioxide and other heat-trapping greenhouse gases, rapidly increasing global temperature, raising sea levels, exacerbating drought and wildfire, intensifying storms and heat waves, and ultimately endangering billions of people and countless nonhuman species. One of the many obstacles to public and political reckoning with climate change has been the stubborn notion that humans are not powerful enough to affect the entire planet. In truth, we are far from the only creatures with such power. The history of life on Earth is the history of life remaking Earth.

AS I STUDIED THE interdependence of Earth and life, I continually returned to an ancient and controversial idea: that Earth itself is alive. Animism is one of humanity's oldest and most ubiquitous beliefs. Throughout history, diverse cultures have extended the concepts of life and spirit to the planet and its components. In many religions, Earth is personified as a deity, often a goddess, who may be maternal, monstrous, or both. The Aztecs worshiped Tlaltecuhtli, a colossal clawed chimera whose dismembered body became the mountains, rivers, and flowers of the world. In Norse mythology, the giantess Jörð's name and identity were synonymous with Earth. Several cultures have imagined Earth as a garden growing from the back of a gargantuan turtle. The ancient Polynesians revered Rangi and Papa, Sky and Earth, who remained locked in a lover's embrace until their children separated them. Still, they cried out for one another, in the form of rising mist and falling rain.

The idea that Earth is alive seeped from myth and religion into early Western science, persisting for centuries. Many ancient Greek philosophers regarded Earth and other planets as animate entities with souls or vital forces. Leonardo da Vinci wrote of the parallels between Earth and the human body, comparing bones to rocks, water to blood, and the tides to breathing. James Hutton, the eighteenth-century Scottish scientist who helped found modern geology, described the planet as a "living world" that possessed a "physiology" and the ability to repair itself. Not long after, German naturalist and explorer Alexander von Humboldt characterized nature as a "living whole" in which organisms were connected by a "net-like intricate fabric."

Hutton and Humboldt were exceptions among their peers, however, especially those who hewed to strict empiricism. By the mid-nineteenth century, even metaphorical descriptions of Earth as a living entity were largely out of vogue in the mahogany halls of European science. Academic disciplines were becoming more specialized and reductionist. Scientists were organizing matter and natural phenomena into increasingly specific categories that further segregated life from nonlife. In tandem, the far-reaching consequences of the Industrial Revolution and the expansion of colonial empires favored language and worldviews based on mechanization, profit, and conquest. The planet was no longer perceived as an immense living being worthy of veneration but rather a body of inanimate resources waiting to be exploited.

It was not until the late twentieth century that the idea of a living planet found one of its most popular and enduring expressions in the canon of Western science: the Gaia hypothesis. Conceived by British scientist and inventor James Lovelock in the 1960s and later developed with American biologist Lynn Margulis, the Gaia hypothesis proposes that all the animate and inanimate elements of Earth are "parts and partners of a vast being who in her entirety has the power to maintain our planet as a fit and comfortable habitat for life."* Viewed through the lens of Gaia, Lovelock wrote, Earth is like a giant redwood tree. Only a few parts of a tree contain living cells, namely the leaves and thin layers of tissue within the trunk, branches, and roots. The vast majority of a mature tree is dead wood. Similarly, the bulk of our

* Lovelock and Margulis continually refined the Gaia hypothesis throughout their careers. Many other researchers developed their own interpretations. For examples of the different versions of Gaia, and a scholarly classification, see the supplementary notes at the end of the book.

planet is inanimate rock, wrapped in a flowering skin of life. Just as strips of living tissue are essential to keep a whole tree alive, Earth's living skin helps sustain a kind of global being.

Although Lovelock was not the first scientist to describe Earth as a living entity, the audacity, expansiveness, and eloquence of his vision provoked a tremendous outpouring of acclaim and derision. Lovelock published his first book on Gaia in 1979 amid a growing environmental movement. His ideas found an enthusiastic audience among the public, but they were not as warmly received by the scientific community. Throughout the next several decades, many scientists criticized and ridiculed the Gaia hypothesis. "I would prefer that the Gaia hypothesis be restricted to its natural habitat of station bookstalls, rather than polluting works of serious scholarship," wrote the evolutionary biologist Graham Bell in one review. Robert May, future president of the Royal Society, declared Lovelock "a holy fool." Microbiologist John Postgate was especially vehement: "Gaia—the Great Earth Mother! The planetary organism! Am I the only biologist to suffer a nasty twitch, a feeling of unreality, when the media invite me yet again to take it seriously?"

Over time, however, scientific opposition to Gaia waned. In his early writing, Lovelock occasionally granted Gaia too much agency, which encouraged the misperception that the living Earth was yearning for some optimal state. Yet the essence of the Gaia hypothesis—the idea that life transforms the planet and is integral to its self-regulating processes—was prescient. Although some researchers still recoil at the mention of Gaia, these truths have become tenets of modern Earth system science, a relatively young field that explicitly studies the living and nonliving components of the planet as an integrated whole.* As

* In 2001, thirty-three years after the conference in Princeton, New Jersey, where Lovelock first spoke publicly about what he would eventually call Gaia, several major international scientific organizations signed the historic Amsterdam Declaration on Global Change, which proclaimed in part, "The Earth System behaves as a single, self-regulating system comprised of physical, chemical, biological and human components, with complex interactions and feedbacks between the component parts."

Earth system scientist Tim Lenton has written, he and his colleagues "now think in terms of the coupled evolution of life and the planet, recognizing that the evolution of life has shaped the planet, changes in the planetary environment have shaped life, and together they can be viewed as one process."

Although the idea that Earth itself is a living entity remains controversial, some scientists embrace it and others are increasingly open to it. "There's no question to me that our planet is alive," says atmospheric scientist Colin Goldblatt. "To me, that is a plain statement of fact." Astrobiologist David Grinspoon has written that Earth is not simply a planet with life on it, but rather a living planet. "Life is not something that happened on Earth but something that happened to Earth," he says. "There is this feedback between the living and nonliving parts of the planet that make the planet very different from what it would otherwise be." Even some of Gaia's fiercest critics have changed their minds. "I've got a confession," evolutionary biologist W. Ford Doolittle wrote in *Aeon* in 2020. "I've warmed to Gaia over the years. I was an early and vociferous objector to Lovelock and Margulis's theory, but these days I've begun to suspect that they might have had a point."

Those who bristle at the notion of a living planet will throw up the usual protestations: Earth cannot be alive because it does not eat, grow, reproduce, or evolve through natural selection like "real" living things. We should remember, however, that there has never been an objective measure nor a precise and universally accepted definition of life. Instead, textbooks typically offer a long list of qualities that presumably distinguish the animate from the inanimate. Yet the two cannot be categorically separated. There are numerous examples of things we consider inanimate that have traits of the living and vice versa. Fire, for example, grows as it consumes fuel, and crystals faithfully replicate their highly organized structures as they expand—but most people do not think of either as alive. Conversely, some organisms, such as brine shrimp and the microanimals called tardigrades, can enter a period of extreme dormancy during which they stop eating, growing, and changing in any way, but are still considered living creatures. Most scientists exclude viruses from the realm of the living because they cannot reproduce and evolve without hijacking cells, yet they do not hesitate to ascribe life to all the parasitic animals and plants that are equally incapable of surviving or multiplying without a host.

Life, then, is more spectral than categorical, more verb than noun. Life is not a distinct class of matter, nor a property of matter, but rather a process—a performance. Life is something matter *does*. Although science has not yet settled on a definition or fundamental explanation of life, many experts in the past century have favored a variation of the following: Life is a system that sustains itself.* Living systems use free energy to maintain an improbably high level of organization in a universe hurtling inescapably toward maximum entropy—toward complete dissolution. Theoretically, any system of matter with sufficient complexity and an adequate supply of energy can participate in the phenomenon we call life. We need to become more comfortable with the idea that life happens at many different scales: at the scale of the cell, the organism, the ecosystem, and—yes—the planet.

Like many living things, Earth absorbs, stores, and transforms energy. Earth has a body with organized structures, membranes, and daily rhythms. From the raw elements of our planet have emerged zillions of biological entities that ceaselessly devour, transfigure, and replenish its rock, water, and air. Such organisms do not simply reside on Earth—they are literal extensions of Earth. Moreover, organisms and their environments are inextricably bonded in reciprocal evolution, often converging upon self-stabilizing processes that favor mutual persistence. Collectively, these processes endow Earth with a kind of planetary physiology: with breath, metabolism, a regulated temperature, and a balanced chemistry. Earth is not a single organism, nor a product of standard Darwinian evolution, but it is nonetheless a genuine living entity—a vast interconnected living system. Earth is as alive as we are.

The scientific community's initial reception of Lovelock's hypothesis might have been less disparaging had he given it a different name.

* Systems scientists define a system as an organized network of components that function as a whole. A system can be as small and simple as a molecule or as large and complex as the cosmos. Some scientists have argued that all forms of life, whether microbe, forest, or planet, are by definition living systems.

On the advice of his friend William Golding, author of *Lord of the Flies,* Lovelock named his global being after the Greek goddess Gaia, the personification of Earth, forever branding his ideas with the scientific taboo of anthropomorphism. Whether Lovelock intended it or not, his chosen namesake gave his hypothesis a maternal face and a certain mysticism, making it an easy target for critics with little tolerance for metaphor and hostility toward anything resembling religion or myth. As we reexamine and reanimate the concept of a living planet for the twenty-first century, perhaps we don't need to appropriate old names or invent new monikers. Our planet is an extraordinary living entity that already has a well-known name. It is a creature called Earth.

AS THE LARGEST AND most complex form of life known to us, Earth is also the most difficult to comprehend. Purely mechanistic metaphors fail to capture the vitality and lushness of our planet. Analogies to animal bodies seem too narrow for a planet whose living matter is mostly plants and microbes. Perhaps there is no perfect metaphor, but in the course of writing this book, I have found one that is both useful and complementary to the concept of a living Earth: music.*

As Lynn Margulis wrote, the animate Earth "is an emergent property of interaction among organisms, the spherical planet on which they reside, and an energy source, the sun." Music, too, is an emergent phenomenon: it cannot be reduced to notes on paper, the shape of an instrument, or the dexterous movements of a musician's hands but instead arises from the interaction of all its constituent parts. When the right sequence of notes is played, when it is combined with other sequences in just the right way, we no longer hear mere sounds—we experience music. Likewise, the living entity we call Earth emerges

* Science has a long and fascinating tradition of musical metaphors. The ancient Greeks perceived the orderly movement of the planets as "the music of the spheres." String theory postulates the existence of "tiny strings whose vibrational patterns orchestrate the evolution of the cosmos." Genomes are often compared to musical instruments and gene expression to songs. And the word *organ* (as in the musical instrument), *organ* (as in a vital body part), and *organism* all share the same root, meaning "work."

from a highly complex set of interactions: the mutual transformation of organisms and their environments.

For the first half billion years of its existence, the planet was a purely geological construct. As the first living creatures adapted to the planet's primordial features and rhythms, they began to play upon them, too, each changing the other. Since then, biology and geology, the animate and inanimate, have been locked in a perpetual and increasingly elaborate duet. Over the eons, despite perennial tumult, Earth and its life forms discovered profound harmonies: they regulated global climate, calibrated the chemistry of the atmosphere and ocean, and kept water, air, and vital nutrients cycling through the planet's many layers. Erupting megavolcanoes, asteroid strikes, shriveling seas, and other unimaginable catastrophes have ravaged the planet many times, overwhelming long-established rhythms with pandemonium. Yet our living planet has consistently demonstrated an astonishing resilience—an ability to revive itself in the wake of devastating calamities and find new forms of ecological consonance.

When we learn to see our species as part of a much larger life form—as members of a planetary ensemble—our responsibility to Earth becomes clearer than ever. Human activity has not simply raised global temperature or "harmed the environment"—it has severely imbalanced the largest living creature known to us, pushing it into a state of crisis. The speed and magnitude of this crisis are so great that, if we do not intervene, Earth will require anywhere from thousands to millions of years to fully recover on its own. In the process, it will become a world unlike any we have known—a world incapable of supporting modern human civilization and the ecosystems on which we currently depend.

Our species is unique in its ability to study the Earth system as a whole and deliberately alter it. But it would be hubris to try and control such an immensely complicated system in its entirety. Instead, we must simultaneously acknowledge our disproportionate influence on the planet and accept the limitations of our abilities. The most essential undertaking is clear: in order to avert the worst possible outcomes of the climate crisis, wealthy industrial and postindustrial nations must lead a global effort to rapidly replace fossil fuels with clean and renewable energy. Earth system science underscores the importance of a complementary approach. Our living planet has evolved many ways to store carbon and regulate climate. Over the past few centuries, the ocean and continents, and the ecosystems within them, have absorbed and sequestered much of humanity's greenhouse gas emissions. By protecting and restoring Earth's forests, grasslands, and wetlands—its undersea meadows, abyssal plains, and reefs—we can amplify their planet-stabilizing processes and preserve ecological synchronies that have developed over eons.

Becoming Earth is an exploration of how life has transformed the planet, a meditation on what it means to say that Earth itself is alive, and a celebration of the wondrous ecology that sustains our world. It's a book about how the planet became Earth as we've known it, how it is rapidly becoming a very different world, and how we—we who are alive at this crucial moment in the planet's history—will ultimately help determine what kind of Earth our descendants inherit for millennia to come. The book's three sections—Rock, Water, and Air—mirror the planet's three main elemental components and its three major spheres—the lithosphere, the hydrosphere, and the atmosphere. Their order further reflects their relative abundance: by mass, Earth contains vastly more rock than water and significantly more water than air. Each section has three chapters, the first of which examines how microbes, Earth's earliest and smallest organisms, altered that layer of the planet. The second chapter in each section focuses on key transformations wrought by succeeding waves of larger, more complex life forms—fungi, plants, and animals—and how those changes depended on the ones that came before. The third reviews how our species has rapidly changed Earth in relatively recent history and investigates how best to reform our relationship with the planet.

We'll begin our journey deep within the crust and work our way out, roaming the continents, immersing ourselves in the planet's liquid expanse, and finally reaching the most ethereal of the three spheres, the envelope of air extending more than six thousand miles above us. Along the way, we'll swim through underwater forests, visit an experimental nature park where animals are reconstructing the landscape, and climb to an observatory halfway between the treetops and clouds. We'll meet a diverse cast of fascinating characters—scientists, artists, and inventors; firefighters, spelunkers, and beachcombers—many of whom have dedicated their lives to studying and protecting our living home. We'll travel back in time to some of the most formative events in Earth's tumultuous 4.54-billion-year history and imagine its many possible futures. And we will learn to recognize life's imprint on every part of the planet today, from the heart of the Amazon rainforest to the soil in your backyard.

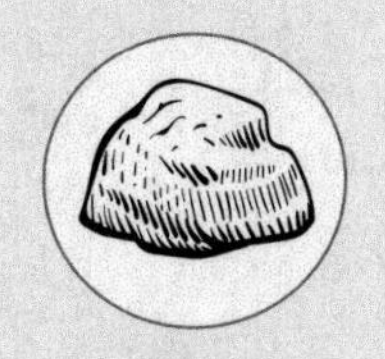

ROCK

CHAPTER

1

INTRATERRESTRIALS

EARTH'S SKIN IS FULL OF PORES, AND EVERY PORE IS A PORTAL TO AN inner world. Some are large enough only for an insect; others could easily accommodate an elephant. Some lead only to minor caves or shallow crevices, whereas others extend into the unexplored recesses of Earth's rocky interior. Any human attempting to journey toward the center of our planet requires a very particular type of passageway: one that is wide enough, yes, but also extremely deep, stable along its entire depth—and ideally equipped with an elevator.

One such portal sits in the middle of North America. About half a mile wide, the furrowed pit spirals 1,250 feet into the ground, exposing a marbled mosaic of young and ancient rock: gray bands of basalt, milky veins of quartz, pale columns of rhyolite, and shimmering constellations of gold. Beneath the pit, some 370 miles of tunnels twist through solid rock, extending more than 1.5 miles below the surface. For 126 years, this site in Lead, South Dakota, housed the biggest, deepest, and most productive gold mine in North America. By the time it ceased operations in the early 2000s, the Homestake Mine had produced more than two million pounds of gold.

In 2006, the Barrick Gold Corporation donated the mine to the state of South Dakota, which ultimately converted it into the largest subterranean laboratory in the United States, the Sanford Underground Research Facility. After mining ceased, the tunnels began to flood. Although the lowest half of the facility remains waterlogged, it is still possible to descend nearly a mile underground. Most of the scientists who do so are physicists conducting highly sensitive experiments that must be shielded from interfering cosmic rays. While the physicists slip into bunny suits and seal themselves in polished laboratories equipped with dark matter detectors, biologists who venture into the underground labyrinth tend to seek out its dankest and dirtiest corners—places where obscure creatures extrude metal and transfigure rock.

On a bitingly cold December morning, I followed three young scientists and a group of Sanford employees into "The Cage"—the bare metal elevator that would take us 4,850 feet into Earth's crust. We wore neon vests, steel-toed boots, hard hats, and, strapped to our belts, personal respirators, which would protect us from carbon monoxide in the event of a fire or explosion. The Cage descended swiftly and surprisingly smoothly, its spare frame revealing glimpses of the mine's many levels. Our idle chatter and laughter were just audible over the din of unspooling cables and whooshing air. After a controlled plummet of about ten minutes, we reached the bottom of the facility.

Our two guides, both former miners, directed us into a pair of small linked rail cars and drove us through a series of narrow tunnels. The cars jostled forward with a sound like the rattling of heavy metal chains as the thin beams of our headlamps illuminated curving walls of dark stone threaded with quartz and specked with silver. Beneath us, I saw flashes of old railing, shallow stands of water, and rocky debris. Although I knew we were deep underground, the tunnels acted like blinders, restricting my perspective to a narrow chute of rock. Glancing at the tunnel's ceiling, I wondered what it would feel like to see the full extent of the planet's crust above us—a pile of rock more than three times as tall as the Empire State Building. Would our depth become palpable the way height does when you peer over the edge of a cliff? Sensing the onset of inverse vertigo, I quickly shifted my gaze straight ahead.

Within twenty minutes, we had traded the relatively cool and well-ventilated region near The Cage for an increasingly hot and muggy corridor. Whereas the surface world was snowy and well below freezing, a mile down—much closer to Earth's geothermal heart—it was about 90°F with nearly 100 percent humidity. Heat seemed to pulse through the rock surrounding us, the air became thick and cloying, and the smell of brimstone seeped into our nostrils. It felt as though we had entered hell's foyer.

The rail cars stopped. We stepped out and walked a short distance to a large plastic spigot protruding from the rock. A pearly stream of water trickled from the wall near the faucet's base, forming rivulets and pools. Hydrogen sulfide wafted from the water—the source of the chamber's odor. Kneeling, I realized that the water was teeming with a stringy white material similar to the skin of a poached egg. Caitlin Casar, a geobiologist, explained that the white fibers were microbes in the genus *Thiothrix,* which join together in long filaments and store sulfur in their cells, giving them a ghostly hue. Here we were, deep within Earth's crust—a place where, without human intervention, there would be no light and little oxygen—yet life was literally gushing from rock. This particular ecological hot spot had earned the nickname "*Thiothrix* Falls."

As I gingerly probed the strands of microbes with a pen, biogeochemist Brittany Kruger opened one of several valves on the spigot before us and began conducting various tests on the discharged fluid. By simply dribbling some of the water into a blue handheld device, reminiscent of a *Star Trek* tricorder, Kruger measured its pH, temperature, and dissolved solids. She clamped filters with extremely tiny pores onto some of the valves to collect any microorganisms drifting through the water. Meanwhile, Casar and environmental engineer Fabrizio Sabba examined a series of rock-filled cartridges that had been hooked up to the spigot. Back at the lab, they would analyze the contents to see if any microbes had flowed into the tubes and survived within them, despite the complete darkness, the lack of nutrients, and the absence of a breathable atmosphere.

On a different level of the mine, we sloshed through mud and shin-high water, stepping carefully to avoid tripping on submerged rails and stray stones. Here and there, delicate white crystals ornamented the ground and walls—most likely gypsum or calcite, the scientists told me. When our headlamps caught the tunnels of pitch-black rock at the right angle, the crystals shimmered like stars. After another twenty-minute journey, this time on foot, we reached another large spigot jutting out from the rock. Only half a mile underground, and better ventilated, this alcove was much cooler than the last. The rock around the faucet was mired in what looked like wet clay, which varied in color from pale salmon to brick red. This, too, Casar explained, was the work of microbes, in this case a genus known as *Gallionella,* which thrives in iron-rich waters and excretes twisted metal spires. At Casar's request, I filled a jug with fracture water from the faucet, scooped microbe-rich mud into plastic tubes, and stored them in coolers, where they would await future analysis.

Kruger and Casar have visited the former Homestake Mine at least twice a year for many years. Every time they return, they encounter enigmatic microbes that have never been successfully grown in a laboratory and species that have not yet been named. Their studies are part of a collaborative effort co-led by Magdalena Osburn, a professor at Northwestern University and a prominent member of the relatively new field known as geomicrobiology.

Osburn and her colleagues have shown that, contrary to long-held assumptions, Earth's interior is not barren. In fact, the majority of the planet's microbes—perhaps more than 90 percent—may live deep underground. These intraterrestrial microbes tend to be quite different from their counterparts on the surface. They are ancient and slow, reproducing infrequently and possibly living for millions of years. They often acquire energy in unusual ways, breathing rock instead of oxygen. And they seem capable of weathering geological cataclysms that would annihilate most creatures. Like the many tiny organisms in the ocean and atmosphere, the unique microbes within Earth's crust do not simply inhabit their surroundings—they transform them. Subsurface microbes carve vast caverns, concentrate minerals and precious metals, and regulate the global cycling of carbon and nutrients. Microbes may even have helped construct the continents, literally laying the groundwork for all other terrestrial life.

THE STORY OF THE living rock we call Earth is a story of perpetual metamorphosis. The world our species has known is only one of the planet's successive and often radically distinct identities. Many of Earth's previous permutations would have been inhospitable and largely unrecognizable not only to humans but to any creature apart from a primordial microbe.

When Earth first formed, it was a seething ball of molten rock, likely too small, hot, and volatile to hold on to liquid water or maintain an atmosphere for long. Whatever incipient atmosphere might have existed was probably obliterated about 4.5 billion years ago in an unfathomably violent collision between Earth and one of its smaller sibling planets. The impact resulted in a massive field of rocky debris, some of which eventually coalesced into the Moon. Over the next 100 million years, Earth's molten surface cooled and formed a crust, expelling steam and other gases, including carbon dioxide, nitrogen, methane, and ammonia. Ongoing volcanic activity thickened this gaseous cloak. A continual barrage of asteroids and meteorites delivered more water vapor, carbon dioxide, and nitrogen on impact.

Together, all the gases released from the planet's interior and supplied by hurtling space rocks created a new atmosphere. Immense volumes of water vapor condensed into clouds and poured back to the surface in heavy rains that may have lasted for millennia. By four billion years ago, if not earlier, the liquid water accumulating on the nascent crust had become a shallow global ocean pockmarked by volcanic islands, which gradually grew into the first landmasses.

Like so much in Earth's earliest history, exactly where and when the planet first stirred to life is not definitively known. At some point not long after our planet's genesis, in some warm, wet pocket with the right chemistry and an adequate flow of free energy—a hot spring, an impact crater, a hydrothermal vent on the ocean floor—bits of Earth rearranged themselves into the first self-replicating entities, which eventually evolved into cells. Evidence from the fossil record and chemical analysis of the oldest rocks ever discovered indicate that microbial life existed at least 3.5 billion years ago and possibly as far back as 4.2 billion years ago.

Among all living creatures, the peculiar microbes that dwell deep within the planet's crust today may most closely resemble some of the earliest single-celled organisms that ever existed. Collectively, these subsurface microbes comprise an estimated 10 to 20 percent of the biomass—all the living matter—on Earth. Yet until the mid-twentieth century, most scientists did not think subterranean life of any kind was plausible below a few meters.

Humans undoubtedly began to encounter the shallowest and most conspicuous forms of underground life as soon as they began to explore and inhabit caves, but the oldest surviving reports of such encounters date only to the 1600s. In 1684, while traveling throughout central Slovenia, naturalist Janez Vajkard Valvasor investigated rumors of a mysterious spring near Ljubljana, beneath which a dragon was thought to live. Locals believed the dragon forced water to the surface every time it shifted its body. After heavy rains, they explained, they sometimes found baby dragons washed up on rocks nearby—slender and sinuous with blunted snouts, frilled throats, and nearly translucent pink skin. Based on these reports, Valvasor described the animals as "akin to a lizard, in short, a worm and vermin of which there are many hereabouts." It was not for another century that naturalists formally identified the creatures as aquatic salamanders that lived exclusively underground in water flowing through limestone caves. They are now known as olms.

In 1793, Alexander von Humboldt published one of his earliest scientific studies: a monograph on fungi, moss, and algae that eked out a living in mines near Saxony, Germany. Nearly four decades later, in September 1831, cave guide and lamplighter Luka Čeč found a tiny copper beetle, less than a third of an inch long, scuttling through caves in southwestern Slovenia. It was somewhat antlike in appearance, with a bulbous abdomen, narrow head, and spindly legs. Upon closer examination, entomologist Ferdinand Schmidt determined that the beetle was a previously unknown species that had adapted to life underground: it had no wings or eyes, instead navigating its surroundings with its long, bristled antennae. The news of this discovery initiated a flurry of scientific exploration. Between 1832 and 1884, naturalists documented numerous cave-dwelling species new to science, including various crickets, pseudoscorpions, woodlice, spiders, millipedes, centipedes, and snails.

In the early twentieth century, scientists started to get glimpses of the true abundance of life deep underground. Around 1910, while trying to determine the source of methane gas in mines, German microbiologists isolated bacteria from coal samples collected 3,600 feet below the surface. In 1911, the Russian scientist V. L. Omelianski discovered viable bacteria preserved in permafrost alongside an unearthed mammoth. Not long after, UC Berkeley soil microbiologist Charles B. Lipman reported that he had revived ancient bacterial spores trapped in chunks of coal obtained from a Pennsylvania mine.

Although tantalizing, these early studies did not persuade most scientists that microbes were prevalent in the deep crust because of the possibility that surface microbes had contaminated the samples. Over the next few decades, however, researchers continued to find microbes in rock and water obtained from mines and drill sites in Asia, Europe, and the Americas. Soviet biologists even started using the term "geological microbiology." By the 1980s, attitudes in the scientific community had started to shift. Studies of aquifers—underground reservoirs of water embedded in rock—indicated that bacteria populated groundwater and changed its chemistry, even thousands of feet below the surface. The United States Department of Energy launched a Subsurface Science Program to monitor groundwater contamination and investigate whether microbes could help filter any pollutants. Program manager Frank J. Wobber and his colleagues developed more rigorous methods for preventing the accidental introduction of surface microbes, such as disinfecting drill bits and rock cores and tracking the movement of fluids through the crust to make sure surface water was not mingling with their samples.

Ultimately, the results of this research and similar studies confirmed that, if anything, early proponents of a subterranean biosphere had been too conservative in their estimates. Wherever scientists looked—within the continental crust, beneath the seafloor, under Antarctic ice—they found unique communities of microbes collectively containing thousands of unidentified species. Sometimes, microbes were clearly present but diffuse: in certain pockets of the crust, there appeared to be as few as one microbe per cubic centimeter, equivalent to a country with only one person every four hundred miles. The underworld was real, but its inhabitants were much smaller and stranger than anyone had imagined.

In the 1990s, Cornell astrophysicist Thomas Gold published a series of provocative claims about the microbial underland. Gold proposed that microorganisms permeated the entire subsurface, living in fluid-filled pores between the grains in rocks, sustaining themselves not with light and oxygen but primarily with methane, hydrogen, and metals. Although scientists had not yet found microbes further than 1.86 miles underground, Gold suspected that they lived even deeper—up to 6.2 miles—and that the biomass within the crust was at least equal to, if not greater than, that on the surface. He further suggested that all life on Earth, or at least some branches of it, may have originated in the planet's interior; that other planets and moons might also harbor subterranean ecosystems; and that deep-dwelling microbes, protected from surface vicissitudes, were likely the most common form of life throughout the cosmos.

By the early 2000s, motivated in part by Gold's vision, scientists had started negotiating new ways to plunge even farther into Earth's crust. Mines were particularly promising because they provided access to the remote subsurface without requiring much additional drilling or infrastructure. Tullis Onstott, a professor of geosciences at Princeton University, and his colleagues traveled to ultra-deep gold mines in South Africa and retrieved samples of groundwater from nearly two miles underground. Within some of the deepest samples, they found a single species: a baguette-shaped bacterium with a whiplike tail that endured temperatures up to 140°F and acquired energy from the chemical byproducts of radioactively decaying uranium embedded in its sunless home.

Onstott and his colleagues decided to name the microbe *Desulforudis audaxviator* after a passage in Jules Verne's *Journey to the Center of the Earth,* which reads "descende, Audax viator, et terrestre centrum attinges"—descend, bold traveler, and you will attain the center of the Earth. The water in which *D. audaxviator* was discovered had not been disturbed for tens of millions of years at a minimum, suggesting that a population of these daring microbial terranauts may have sustained itself for at least as long. "We do not normally think of rock as harboring life," Onstott writes in his book *Deep Life*. "I am a geologist by training, and like most geologists, I too have viewed rocks as inanimate entities." But now, he continues, as a geomicrobiologist, he sees all rocks as little worlds unto themselves, composed of microorganisms, "some of which may have been living in the rock since its formation hundreds of millions of years ago."

Some communities of subsurface microbes may be even older. The Kidd Creek Mine in Ontario, Canada, is one of the largest and deepest mines in the world: extending about 1.86 miles belowground, it contains rich veins of copper, silver, and zinc that formed nearly three billion years ago on the ocean floor. In 2013, University of Toronto geologist Barbara Sherwood Lollar published a study demonstrating that some parcels of water in the Kidd Creek Mine have been untouched and isolated from the surface for more than a billion years—the oldest water ever discovered on Earth. Transparent when first collected, the iron-rich water tinges a pale orange when exposed to oxygen; it has the consistency of light maple syrup, contains at least twice as much salt as modern seawater, and, in Sherwood Lollar's judgment, "tastes terrible." In 2019, Sherwood Lollar, Magdalena Osburn, and several colleagues confirmed that, just like much younger fluids circulating through the pores and fissures in rock a few thousand feet below the surface, the billion-year-old water miles deep in Kidd Creek Mine is also populated by microorganisms. Like many deep Earth microbes, they too depend on the molecular byproducts of radiation-powered chemical reactions between rock and water. Whether some of these microbes are themselves an eon old is not yet known, but it is plausible.

"This research really is a form of exploration," Sherwood Lollar says. "Some of the findings are causing us to rewrite the textbooks about how this planet works. They are changing our understanding of Earth's habitability. We don't know where life originated. We don't know if life arose on the surface and went down, or whether life emerged below and went up. There's a tendency to think about Darwin's warm little pond, but, as my colleague T. C. Onstott likes to say, it could just as easily have been some warm little fracture."

Even close to two miles deep in Earth's bowels, there is more than microbial life. Scientists have found fungi, flatworms, arthropods, and microscopic aquatic animals known as rotifers living in South African gold mines between 0.62 and 1.86 miles below the surface. In December 2008, Onstott's colleague Gaëtan Borgonie, a Belgian zoologist, made a discovery uncannily reminiscent of the "vermin" Valvasor described in the seventeenth century. About 0.8 miles deep in Beatrix Gold Mine, near the town of Welkom, South Africa, he collected a nematode—a tiny roundworm—from filtered borehole water. At low magnification, it looked like nothing more than a squirming noodle half a millimeter long, about five hundred times larger than a bacterium. Under a powerful scanning electron microscope, it resembled a plump leech whose face was wreathed with mouth plates and sensory buds.

In the lab, Borgonie found that the nematode favored a diet of subsurface microbes over more typical roundworm fare. Eventually, it produced twelve eggs, all of which hatched and established a new population. Although the nematode's ancestors had almost certainly washed down into the subsurface with rain, rather than originating there, it was clearly adapted to subterranean life. Borgonie, Onstott, and their colleague Derek Litthauer named the new species *Halicephalobus mephisto* after Mephistopheles, the Devil's delegate in the legend of Faust. It remains one of the most astonishing discoveries in the history of biology. Finding a multicellular animal of that size and complexity living in a trickle of water so deep within the planet's crust was,

Onstott said, like "finding Moby Dick swimming around in Lake Ontario."

MAGDALENA OSBURN'S SHELVES ARE full of rocks, and every rock is the fossil of a story. When we met in her office at Northwestern University, she showed me the Hawaiian basalt she had scooped up with a stick when it was still flowing lava; the giant quartz crystal she had pulled out of a fracture on a trip to Hot Springs, Arkansas; and the pyrrhotite she had surreptitiously stuffed down her overalls while touring a mine in Canada. Near her desk she kept a rippled segment of a 580-million-year-old microbial mat and a milky blue mineral known as smithsonite, which was mined in Magdalena, New Mexico, her namesake. In a different corner of her office, she handed me an orange rock with the texture of sesame brittle. "Those are ooids," she said. "They're like tiny marine gobstoppers of carbonate. If you go to the Bahamas and play around in the shoals, those are largely ooids." She picked up a large gray chunk of amphibolite: "This rock tried to kill me. I was in a rock fall at a field camp when I was an undergrad. This rock actually went through my tent." "So it just missed you?" I asked. "Well, I was running like hell in the other direction," she explained. "And when I got back to my tent, this rock was there. So that's the death rock."

Rocks and the stories they tell have been an important part of Osburn's life since childhood. Her father was a laboratory administrator in the Department of Earth and Planetary Sciences at Washington University in St. Louis. She often joined him on university field trips to see bluffs, gorges, billion-year-old lava rock, massive boulders deposited by glaciers, and other geological features of Missouri. "It would always be a bunch of Wash U. students and then me as a seven-year-old or something," she recalled. "And I was always the one that was too close to the cliff, or too high up the cliff, or with my head hanging off." On one occasion she slashed her hand open on a rock and started gushing blood. As the college students gaped in horror, she sauntered up to her father and casually requested a bandage.

Many of Osburn's formative experiences with scientific research focused on the intersection of geology and microbiology. As an undergrad at Washington University, she studied hot springs and shallow hydrothermal vents, both sites where certain heat-loving bacteria thrive. At the California Institute of Technology, her graduate thesis combined studies of ancient rock with innovative analyses of the chemical signatures that modern microbes leave in their environment, ultimately working toward a new understanding of how microbes have changed Earth throughout its history.

While completing a postdoc at the University of Southern California, Osburn became one of the lead investigators on a NASA-funded study of subsurface life and visited the former Homestake Mine for the first time. In the past, miners searching for high-quality ore had drilled exploratory boreholes, some of which tapped into underground reservoirs of water. After removing the cores for analysis, miners had plugged the holes with concrete, although some of them continued to leak. When Osburn's team discovered that several of these leaky conduits contained microbes, they arranged for the miners to clear out the boreholes with a diamond-bladed industrial drill. They then replaced the concrete with plastic tubes fitted with valves so that they could periodically return and collect new samples, establishing a functional underground observatory.

After admiring the rock collection in her office, Osburn and I walked over to her microbiology lab, where she stores water, sediment, and microbes collected from various research sites. Osburn and Caitlin Casar prepared several microscope slides, smearing some mine water across the glass. Osburn removed her tortoiseshell glasses and tucked her wavy brown hair out of the way before situating herself in front of the microscope and adjusting various knobs to get a clear view. "Here we are looking at some *Gallionella* with twisty, twisty stalks," she said. At about 1000x magnification, the microbes looked like daubs of orange marmalade and caviar. On the computer screen linked to the scope it was easier to see the stalks Osburn had referenced: contorted filaments of iron, some like warped corkscrews, others like loosely braided twist ties—byproducts of the microbes' unique metabolism. A few minutes later we switched to looking at *Thiothrix,* which resembled white baubles tangled in yellow tinsel. We could see the bright dots of sulfur compounds that the microbes had sequestered within their cells when they converted the element from one molecular state to another.

I thought I saw something moving and asked what it might be. "This is a pretty old sample of biofilm," Osburn said, "so I doubt there's much—Oh!" A minuscule dot twitched across the screen like a jumping bean. "Just as I said it was dead, there's a happy little cell." The sample we were examining had been collected several years earlier and, as it was never intended for culturing cells, it had not received any special care. Yet, somehow, this drop of rock and water—this dribble from Earth's deepest veins—was still pulsing with life.

FOR HUNDREDS OF YEARS, Lechuguilla Cave appeared to be little more than a long hole leading to dead-end passages in the Guadalupe Mountains of New Mexico. Explorers occasionally ventured into the pit. Prospectors routinely visited to collect bat guano, which was prized as a fertilizer. Otherwise, no one paid it much attention. One blustery day in the 1950s, however, cavers noticed air streaming through rubble at the bottom of the cave, suggesting that there might be a hidden section. A series of excavations in the 1970s and '80s uncovered several long passageways. Subsequent explorations eventually revealed more than 145 miles of underground terrain extending more than 1,600 feet below the surface. The tunnels and chambers were decorated with strange and beautiful formations: massive chandeliers of frostlike gypsum, lemon-yellow sulfur pods, pearly balloons of hydromagnesite, transparent selenite spears, and calcite lily pads hovering over turquoise pools.

In the early 1990s, microbiologist Penny Boston watched a *National Geographic* TV special that mentioned Lechuguilla. She was fascinated by the prospect of a pristine subterranean wonderland. One of the researchers featured on the show, Kim Cunningham, had found some preliminary evidence of microbial life in the cave. Boston, who was particularly interested in the possibility of life beyond Earth, saw Lechuguilla as an analog for potential subsurface habitats on other planets. She called Cunningham and arranged to visit the cave with a team of scientists and cavers.

Boston and the other scientists, who did not have much caving experience, practiced for a few hours on cliffs in Boulder, Colorado, before taking the plunge into Lechuguilla. The brief training was nowhere near adequate. Lechuguilla is not a simple series of interconnected rooms that one can walk through; it's a tangle of crystal labyrinths embedded in a tortuous maze of rock. To navigate it, Boston and her colleagues had to rappel down steep cliffs, climb over slippery towers of gypsum, traverse narrow ledges of rock, and squirm through stone honeycombs—all while towing their cumbersome gear. "We were in such an alien environment that we were just basically coping," Boston recalls. "I kept thinking to myself, I just have to live long enough to get out of here."

They did survive, but not without injuries. At one point, Boston sprained her ankle. While shimmying across a crevasse, she gouged her shin, which caused her leg and foot to swell. She pressed on. Not long before leaving the cave, she spied some curious rust-colored fluff coating a low-hanging portion of the ceiling. She was preparing to scrape some of the fluff into a bag when a smidgen fell into her eye, which soon swelled shut as though it were infected. Perhaps, she thought, the brown fuzz was made by microbes; perhaps it was made *of* microbes. Laboratory studies eventually confirmed her hunch: the cave was covered in microorganisms that chewed through rock, extracting iron and manganese for energy and leaving behind a soft mineral residue. Microbes were turning rock into soil more than one thousand feet underground.

Eventually, through many years of research, Boston and other scientists—including Diana Northup, Carol Hill, and Jennifer Macalady—revealed that the microbes in Lechuguilla do much more than spit out a little dirt. Lechuguilla is ensconced in thick layers of limestone, the petrified remains of a 250-million-year-old reef. The manifold chambers in such caves are usually formed by rainwater that seeps into the ground and gradually dissolves the limestone. In Lechuguilla, however, microbes are the sculptors: bacteria eating buried reserves of oil release hydrogen sulfide gas, which reacts with oxygen in groundwater, producing sulfuric acid that carves away limestone. In parallel, different microbes consume hydrogen sulfide and generate sulfuric acid as a byproduct. Similar processes happen in 5 to 10 percent of limestone caverns globally. Although such caves could form from the purely geological production of acids and gases, microbes amplify the process, allowing the chambers to grow much larger much faster.

Since Boston's initial descent into Lechuguilla, scientists around the world have discovered that microorganisms transform the planet's crust wherever they inhabit it—which is to say, just about everywhere. Alexis Templeton, a geomicrobiologist at the University of Colorado, Boulder, regularly visits a barren mountain valley in Oman where tectonic activity has pushed sections of the earth's mantle—the layer that sits below the crust—much closer to the surface. She and her colleagues drill a borehole up to a quarter of a mile into the uplifted mantle and extract long cylinders of 80-million-year-old rock, some of which are beautifully marbled in striking shades of maroon and green. In laboratory studies, Templeton has demonstrated that these samples are full of bacteria that change the composition of Earth's crust: they eat hydrogen and breathe sulfates in the rock, exhale hydrogen sulfide, and create new deposits of sulfide minerals similar to pyrite, also known as fool's gold.

Through related processes, microbes have helped form some of Earth's deposits of gold, silver, iron, copper, lead, and zinc, among other metals. As subsurface microbes break down rock, they often free the metals stuck within it. Some of the chemicals microbes release, such as hydrogen sulfide, combine with free-floating metals, forming new solid compounds. Other molecules produced by microbes grab soluble metals and bind them together. Some microbes stockpile metal inside their cells or grow a crust of microscopic metal flakes that continuously attracts even more metal, potentially forming a substantial deposit over long periods of time.

Life, in particular microbial life, has also forged the majority of

Earth's minerals, which are naturally occurring inorganic solid compounds with highly organized atomic structures—or, to put it more plainly, very elegant rocks. Like living organisms, minerals are classified into families and species. Today, Earth has at least five thousand distinct mineral species, most of which are crystals such as diamond, quartz, topaz, graphite, and calcite. In its infancy, however, Earth did not have much mineral diversity. Over time, the continuous crumbling, melting, and resolidifying of the planet's early crust shifted and concentrated uncommon elements. Life began to break apart rock and recycle elements, generating entirely new chemical processes of mineralization. More than half of all minerals on the planet can occur only in a high-oxygen environment, which did not exist before microbes, algae, and plants oxygenated the ocean and atmosphere.

Through the combination of tectonic activity and the ceaseless bustle of life, Earth developed a mineral repertoire unmatched by any other known planetary body. Comparatively, the Moon, Mercury, and Mars are minerally impoverished, with perhaps a few hundred mineral species between them at most. The variety of minerals on Earth depends not merely on the existence of life but also on its idiosyncrasies. Robert Hazen, an Earth scientist at the Carnegie Institution, and statistician Grethe Hystad have calculated that the chance of two planets having an identical set of mineral species is 1 in 10^{322}. Given that there are only an estimated 10^{25} Earthlike planets in the cosmos, there is almost certainly no other planet with Earth's exact complement of minerals. "The realization that Earth's mineral evolution depends so directly on biological evolution is somewhat shocking," writes Hazen. "It represents a fundamental shift from the viewpoint of a few decades ago, when my mineralogy PhD adviser told me, 'Don't take a biology course. You'll never use it!'"

The microbial modification of the crust is not restricted to land—it happens within and beneath the seafloor, too. In some regions, microbes in the ocean crust convert sulfur into sulfate; because sulfate is water-soluble, it dissolves into the sea and becomes an accessible nutrient for other creatures. Marine sediments contain one of the largest reservoirs of methane on the planet, 80 percent of which is produced by microbes. Were all that methane to rise into the atmosphere, it would significantly thicken Earth's invisible blanket of heat-trapping greenhouse gases and greatly intensify global warming. But another set of microbes recycles 90 percent of the methane rising through seafloor sediments before it reaches the surface, constituting "one of the most important controls on greenhouse gas emission and climate on Earth," as one group of experts on deep ocean microbiology has called it.

The continents themselves may also be partial constructs of microbial terraforming. No one knows precisely how the continents were born, but a widely supported theory proposes that continental crust is a distillation of oceanic crust. The continents are made of granite, which, as far as we know, is abundant only on Earth—it has rarely been found anywhere else in the universe. In contrast, oceanic crust is composed of basalt, a cosmically common rock. Basalt is dark, dense, and rich in magnesium and iron, a particularly heavy metal. More than four billion years ago, as Earth's earliest ocean crust aged and cooled, it eventually became heavier than the mantle on which it floated and started to sink—a process called subduction. During its descent into the mantle, ocean crust and its overlying sedimentary layer released the water trapped within them, which lowered the melting point of the surrounding mantle. Certain components of the mantle began to melt into buoyant magma, which eventually erupted from volcanoes and cooled into new rock.

This process continues today. In Earth's earliest chapters, however, the mantle was significantly hotter than it is now; in addition to squeezing water out of sinking ocean crust, the mantle melted the crust itself. When this hybrid magma rose to the surface, it cooled into a new kind of rock—granitoid rock—which was largely depleted of magnesium and iron and thus was much less dense than basalt. Over time, granitoid rock was subducted and recycled into true granite. Because granite was less dense than basalt, it accumulated on top of the ocean crust, forming thick patches of early continental crust that gradually breached the water's surface. Later, with the emergence of plate tectonics, protocontinents coalesced into microcontinents and eventually formed immense tracts of land high above sea level. By about

2.5 billion years ago, nearly a third of the planet's surface was land, a proportion that has fluctuated throughout Earth history with the rising and falling of the seas.

Several Earth scientists, including Robert Hazen and his colleagues, have investigated the possibility that life helped create the continents by promoting the subduction of oceanic crust and sediments and their transformation into granite. The more water the crust and sediments contain, the more easily this process occurs. When Earth was young, microbes inhabiting the ocean crust probably dissolved the basalt with acids and enzymes in order to obtain energy and nutrients, producing wet clay minerals as byproducts and thereby effectively lubricating the crust. A more hydrated crust would have introduced more water into the mantle, accelerating the dissolution of both mantle and crust and their eventual transfiguration into new land.

Geophysicists Dennis Höning and Tilman Spohn have published similar ideas. They point out that water trapped in subducting sediments escapes first, whereas water in the crust is typically expelled at greater depths. The thicker the sedimentary layer covering the crust, the more water makes it into the deep mantle, which ultimately enhances the production of granite. In Earth's earliest eons, microorganisms—and, later, fungi and plants—continuously dissolved and degraded rock at a rate much greater than geological processes could accomplish on their own. In doing so, they would have increased the amount of sediment deposited in deep ocean trenches, thereby cloaking subducting plates of ocean crust in thicker protective layers, flushing more water into the mantle, and ultimately contributing to the creation of new land. Computer models suggest that had life never evolved, the expansion of the continents would have been severely stunted and the planet might have remained a water world flecked with islands—an Earth without much earth.

ABOUT SEVENTY MILES SOUTH of the former Homestake Mine, in a bed of limestone surrounded by rolling prairie, there's a hole in the shape of a heart. Sometimes you can hear it sighing. In English, it's known as

Wind Cave. To the Lakota, it is Maka Oniye—"breathing earth"—and it is sacred. The Lakota regard the Black Hills, the ancient mountainous terrain encompassing the mine and cave, as the womb of Mother Earth. The region is at the center of an ongoing legal dispute with the U.S. government. The 1868 Treaty of Fort Laramie formally established the Lakota as the owners of the Black Hills and protected the area from white settlement. In the 1870s, however, when soldiers and prospectors confirmed rumors of gold in various parts of the Black Hills, the U.S. government revoked the earlier treaty and seized the land.

As the Lakota tell it, Maka Oniye conceals a portal that links the spirit world to the surface of Earth. Long ago, their ancestors lived in the spirit lodge, waiting for the Creator to make the surface habitable. In some versions of the story, a duplicitous spirit lures a group of people through the portal to the surface before it is ready. As punishment, the Creator transforms them into the first bison. Once plants and animals are abundant aboveground, the rest of the people in the spirit land emerge and thrive.

Underground realms are prevalent in religion and literature around the world: the Greek Hades, Hindu Patala, Inuit Adlivun, Aztec Mictlan, and Christian hell; the thousands of legends, folktales, and novels about primeval landscapes, fantastical beasts, magical beings, and advanced civilizations hidden far beneath the surface. In the Americas, origin stories known as earth-diver myths are particularly common. In these myths, a creator, hero, or council of beings asks various animals—beavers, birds, crustaceans—to plunge into primordial waters and retrieve a bit of mud or dirt from which to construct the continents. "The earth was all water," begins one such tale from the Eufaula people of southeastern North America. "Men, animals, and all insects and created beings met and agreed to adopt some plan to enable them to inhabit the earth. They understood that beneath the water there was earth, and the problem to be solved was how to get the earth to the top and spread it that it might become habitable."

Even millions of years ago, early humans must have realized that there was as much to explore below the ground as above. Earth betrays its underland in myriad ways. An old oak tumbles, tipping its buried branches—its secret mirror image—into the air. Someone slips in a limestone cave, falling into a hidden passageway. The ground trembles, ripples, and fractures, opening a depthless trench. That life depends on, and often arises from, the underworld would have been apparent long ago, too. Seedlings suspend themselves in soil, lifting bowed heads toward the sun as they thrust roots ever deeper. Mushrooms materialize from dirt overnight, then collapse back into it almost as quickly. Beetles squirm out of interred cocoons. Bears lumber forth from the darkness of their dens. Humans have been burying their dead for at least eighty thousand years, and probably far longer. When an individual's life ends, our instinct has long been to return them to the earth—to the womb that bore us all.

Science redefines the contour of the plausible: what was once thought certain may dissolve into absurdity as the formerly ludicrous becomes credible. It is easy enough to imagine life in the deepest strata of soil, where there is still a supply of air and nutrients. To accept as scientific fact that life extends much farther—that it pervades the bleak, torrid, crushing expanse of crust miles below the surface—requires extraordinary evidence. Yet this is exactly what scientists over the past four decades have confirmed.

To recognize that deep subsurface life not only exists but is also engaged in a continuous alchemy of earth—that it may have helped create the very crust it inhabits and on which all terrestrial life stands—is to redefine the modern understanding of how our planet came to be. Yet it is also an echo of an ancient truth, one that seems to have lurked in human consciousness for millennia, waiting to be fully unearthed. In the Eufaula's earth-diver myth, the continents do not simply grow out of ocean mud; they must be sculpted. The assembled beings eventually select Crawfish to search for a foundational piece of land. He "went down and after a long time brought up in his claws a ball of earth. This was kneaded, manipulated, and spread over the waters (the great deep). Thus the land was formed."

CHAPTER

2

THE MAMMOTH STEPPE AND THE ELEPHANT'S FOOTPRINT

FROM A DISTANCE IT LOOKED LIKE SNOW: SOFT WHITE CLUMPS scattered along the slopes and shores of Wrangel Island. Even now, in late summer, it was not unusual to encounter ice in these remote Arctic waters, but snow at low elevations was surprising. As he got closer, Nikita Zimov became even less certain about the scene before him. The snow seemed to move. Gradually, each clump began to take on a more distinct shape. They had hummocked backs and thick legs, small black eyes and large round noses. They were, Nikita realized, polar bears. Numerous fully grown polar bears.

Nikita, who was twenty-six at the time, had traveled to the island on an old gray boat with his father, Sergey Zimov, a renowned Arctic ecologist, along with Victor Sorokovikov, a family friend and soil scientist, and Alexey Tretyakov, a young man from their hometown who had agreed to help. The journey took seven days. They departed from Chersky, a tiny Siberian settlement where the Zimovs managed a research station, traveling seventy miles by road along the Kolyma River toward the frigid sea and several hundred more miles by boat. With the exception of some moderately rough waves and a failed transformer, the first few days of the voyage were surprisingly uneventful. They were prepared for the possibility of a storm or a collision with an iceberg in the night, but nothing of the sort happened. They took turns navigating and scanning for obstacles. They ate ham, eggs, ramen, dumplings, and borscht accompanied by generous pours of beer and vodka. When they weren't working, they read, played cards, watched movies, and slept.

On the fourth day, however, as they turned away from the coast to make a final dash through the open sea toward their destination, they encountered a massive ice floe. Circumventing it added another three days to the trip. They spent the nights anchored to inlets of ice, sometimes sleeping alongside walruses with bristled lips and three-foot tusks that arced from their mouths like ivory sabers.

When the group finally arrived at Wrangel Island, they found a safe spot to dock far from the polar bears and stepped onto land for the first time in a week. Shipping crates, decaying boats, and rusty oil barrels littered the shoreline. A giant tree stump lay battered and broken, having repeatedly been hacked for firewood. Nail-studded boards barricaded the windows of the few small shacks and houses. A biologist named Natasha straddled the roof of a nearby house, scanning the sea ice through binoculars in search of walruses, a can of bear spray belted to her hip. The man the Zimovs had come to meet was in the next town over, but evening was rapidly approaching. After enjoying some vodka and the island's makeshift sauna, they retreated to their boat and slept until morning.

Lying 90 miles off the coast of Russia, Wrangel Island is a toupée-shaped blob of tundra almost as large as Yellowstone National Park. Its surface is mostly covered with sand, gravel, and ice. Average temperatures stay well below freezing between December and March and rarely exceed 50°F in summer. Polar winds frequently whip along the island; thick mists occasionally envelop it. Despite these harsh conditions, life is far more abundant on Wrangel than in many other regions of the Arctic. During the last glacial period, smothering ice sheets bypassed the island, which made it something of a refuge. Mammoths survived there for six thousand years after all other members of their species had died out, by which point humans were well into the Bronze Age and the pyramids of Giza had stood for nearly a millennium. Russia has protected the island as a federal nature reserve since the 1970s.

Today, more than four hundred plant species and subspecies take root in Wrangel's soils, twice as many as anywhere else in the Arctic tundra. Moss and lichen carpet whatever ice-free ground they can find. Horned puffins, Pacific loons, and peregrine falcons routinely visit the island, which is also home to the only breeding colony of snow geese in Asia. Arctic foxes and wolves stalk lemmings and reindeer; herds of shaggy muskoxen roam the hills; gray whales and belugas shimmer offshore. Most famously, Wrangel Island is thought to have the highest density of polar bear dens in the world, attracting groups of tourists every year. Other than that, essentially no one is allowed to stay on the island except to conduct scientific research. A few rangers reside there year-round. In summer, a dozen or so researchers might visit, typically staying in an old meteorological station.

Like many scientists, the Zimovs had come to Wrangel Island because of its wildlife. The reserve's director, Alexander Gruzdev, had promised to give them at least six baby muskoxen. The Zimovs planned to relocate the animals to Chersky to take part in a daring and ambitious scientific project—one that, if their theories were correct, would transform vast tracts of the Arctic landscape and help stabilize Earth's climate. On the second morning after their arrival, while exploring the island's various small encampments, they finally met up with Gruzdev. Shortly after introducing themselves and glimpsing the paddock holding the muskoxen, they were interrupted by the sound of an imminent invasion. Three large black dinghies were speeding toward shore, each carrying about twenty elderly American tourists. Apparently, a cruise ship from Alaska had anchored nearby so that its passengers could see the polar bears and puffins up close. "You know, I felt like a great explorer of the Arctic when we finally got to the island," Nikita recalls. But the sight of so many senior citizens reaching the island with such ease, wobbling onto shore with digital cameras in hand, "nullified the experience."

Gruzdev had previously agreed to help guide the tourists around the island, so he temporarily left the Zimovs to occupy themselves. The Zimovs were still hoping to load their living cargo onto their boat and leave fairly soon. The island had different plans. Later that day, when the Zimovs were resting and neither Gruzdev nor the rangers were in town, a polar bear bashed its way through the wood-and-chicken-wire paddock surrounding seven baby muskoxen. As the bear killed one muskox, the other six fled—and with them, the entire reason for the Zimovs' perilous voyage.

NOW IN HIS EARLY FORTIES, Nikita Zimov is tall and lean with glacier-blue eyes, a boyish mop of brown hair, and a small scar on the right side of his chin. His sentences, delivered in a strong Russian accent, tend to swell and swerve like rivers. He likes to try out English slang, and he frequently deploys deadpan humor. ("Of course you need to avoid all icebergs," he told me when recollecting the trip to Wrangel Island. "Or else you can become *Titanic* number two or three. Smaller scale, but still very sad.") Sergey, who is approaching seventy, is a bearish man with long grizzled hair and a bushy beard. His forehead is creased and his lower eyelids pouchy. He smokes constantly and drinks vodka with most meals. He speaks slowly and ponderously in broken English, often slipping into bouts of scientific soliloquy. Together, the Zimovs and their spouses run the Northeast Science Station in Chersky, one of the largest and most important centers of science in the Arctic. They host hundreds of researchers from around the world every year.

When I arrived in Chersky in mid-July, Nikita's wife, Anastasia, met me at the airport and drove the short distance through town to the research station, which is perched on the upper banks of the Kolyma River. The main building contains numerous bedrooms for visitors and an octagonal common room with several long dining tables, a wood stove, and a plush brown couch draped with a bearskin. A massive satellite dish sits atop the residential building, though it is no longer in use. Dirt roads lined with willow and fireweed wind past a small, green-roofed chapel, stacks of shipping crates, and several houses where the Zimovs live. As I toured the site, I could see miles of Arctic terrain stretching beyond the river: a mostly flat patchwork of dense shrubland, scattered conifers, and lakes formed by the thawing of frozen soil.

Sergey Zimov first traveled to the high Arctic as a college student in the 1970s. He intended to study the Arctic's paleogeography—its ancient landscapes—by chemically analyzing samples of water, ice, and soil. As he worked, he became fascinated by the unexpected abundance of bones. When Sergey surveyed his research site and the surrounding area, he rarely saw wildlife: there might be the occasional reindeer, wolf, or migrating bird, but other than that the land around him was barren. Yet, wherever he dug, wherever a hillside crumbled or a river washed away layers of sediment, he found the bones of long-dead animals and bygone species. The ground was a graveyard full of mammoths with spiraling tusks, humpbacked bison, maneless cave lions, elk with 80-pound antlers, and an extinct species of rhinoceros nicknamed "the Siberian unicorn" for the single giant horn on the top of its head. Sometimes he would even find a tuft of fur or a scrap of leathery hide. Technically, these remains were not fossils in the same way as petrified dinosaur bones in a museum. They had not turned to stone. Rather, these were the actual bones, hair, and tissues of ancient animals, preserved in frozen soil for tens of thousands of years. They were proof that the Arctic was once as replete with animals as Africa's savannas.

Geologists call the period of Earth's history spanning from approximately 2.6 million years ago to twelve thousand years ago the Pleistocene. For the last hundred thousand years of that epoch, extensive grasslands ringed the northerly latitudes of the globe. These nearly continuous prairies, known as the mammoth steppe, comprised one of the largest and most productive ecosystems that has ever existed, possibly occupying as much as 40 percent of the world's landmass. The mammoth steppe supported incredible numbers of megafauna: giant herbivores, such as mammoths, mastodons, rhinoceroses, and bison, each weighing more than a ton, as well as numerous hulking predators, including bears, lions, and dire wolves. Smaller and more familiar grazers also roamed the pastures: horses, reindeer, muskoxen, and sheep. When the Pleistocene ended, the grasslands and nearly all their colossal inhabitants disappeared. Sergey had been vaguely aware of this ancient ecosystem on his first visit to the Arctic, but seeing the ubiquity of megafauna bones for himself—touching the tusks of mammoths and feeling the fur of extinct bison—sparked an obsession that has never faded. One question in particular haunted him: What happened?

One of the prevailing explanations was climate change. About twenty thousand years ago, the global climate began to warm, and glaciers that had previously covered much of the planet started melting. Forests of birch and conifer replaced the grass and creeping willow that had nourished most herbivores. As a consequence, scientists had proposed, many cold-adapted, grass-loving species died out. To Sergey, this explanation did not make sense. Around the globe, gargantuan mammals on land and in the sea had survived hundreds of millions of years of repeated glaciations and thaws. Why would they suddenly expire during this one episode of warming in this long-running pattern? Sergey favored a different cause: humans. By about half a million years ago, and possibly much earlier, humans had developed sufficient intelligence and technology to kill some of the largest animals around, such as elks, rhinoceroses, even mammoths—at least occasionally. Humans were gnats in a world of titans, but their ingenuity and manual dexterity, combined with strategic cooperation, eventually made them superpredators. As humans ascended, Sergey reasoned, the planet's giants dwindled.

Geoscientist Paul Martin and climatologist Mikhail Budyko had already published similar ideas in the late 1960s, inciting a debate that continues to this day. In the past decade, however, the argument that humans hunted Pleistocene megafauna into extinction has gained considerable support. Recent evidence from the fossil record and archaeological digs has revealed that wherever humans migrated during the Pleistocene—wherever they brought their spears, arrows, and packs of dogs—large mammals promptly went extinct. As humans spread throughout Europe and Asia roughly 50,000 years ago, dozens of giant mammal species were extinguished. Shortly after humans reached Australia's shores 45,000 years ago, another twenty species of large herbivores disappeared. When humans populated the Americas between 15,000 and 7,000 years ago, more than eighty species weighing at least one hundred pounds each vanished. Although climate change and unstable population dynamics may partly explain this mass eradication, humans were likely the major culprit.

Sergey's most important insight was that such a widespread extinction of megafauna would inevitably have serious ecological repercussions for the planet as a whole. Whereas many of his peers focused on how a shifting climate would have endangered Ice Age creatures, Sergey realized that the causality might flow in both directions. He surmised that mammoths and other colossal mammals had actively maintained their grassland habitat, which had in turn preserved a relatively cool climate. By hunting so many large mammals to extinction, humans might have triggered, or at least exacerbated, the global warming that ended the most recent glaciation. For this reason, some scholars have argued that the Anthropocene—the proposed geologic epoch defined by humanity's profound influence on the planet—should begin somewhere between 50,000 and 10,000 years ago, coinciding with the brunt of megafaunal extinctions.

Sergey's grand theory has its roots in a simple observation: grass is not like most plants. Many plant species protect themselves from herbivores with thick bark, thorns, or toxic or unpalatable chemicals. Some simply grow out of reach. When grasses emerged between 100 and 70 million years ago, some of them evolved a rather different strategy. Sergey and other scientists have proposed that, rather than rely on robust and elaborate defenses, certain grasses negotiated a symbiosis—an ecological partnership—with large herbivores. These grasses offered grazers endless fields of tender green leaves that quickly regenerated when shorn. In exchange for this perpetual sustenance, mammoths and other megafauna trampled, ate, and otherwise deterred the grasses' main botanical competitors, such as shrubs and trees, and fertilized the fields with their copious dung. Together, the theory goes, grasses and megafauna created and regulated the mammoth steppe ecosystem.

The more Sergey thought about this symbiosis, the more aware of its power he became. The alliance between grasses and grazers would have changed far more than their local landscapes. In the Pleistocene, as today, thick layers of frozen soil known as permafrost lay beneath the Arctic's surface, concealing a vast reservoir of carbon in the form of the preserved remains of ancient life. If the climate changed and the temperature rose high enough, the permafrost began to thaw, allowing microbes to break down its organic matter and potentially releasing potent greenhouse gases such as carbon dioxide and methane in sufficient quantities to warm the planet. The mammoth steppe, Sergey reasoned, would have counteracted this global warming. Because grasses were typically paler than trees and many other plants, they had a much higher albedo, or reflectivity, bouncing more light back into space and thus cooling Earth. Grasses also captured enormous amounts of carbon from the atmosphere and stored it in deep, diffuse roots while simultaneously sponging up most of the water in the ground, keeping it dry, firm, and intact. In the winter, mammoths and other large herbivores stripped away superficial, heat-trapping layers of snow through the simple act of walking on the ground with their considerable heft as well as by digging through the snow to uncover buried plants and tubers. In doing so, they exposed the permafrost to frigid winter temperatures, assuring that it remained frozen. When humans killed off most of the world's megafauna, they did much more than reduce Earth's biodiversity—they also impaired its ability to regulate global climate.

The complex ecological bonds between grazers, grasslands, and climate emerge from one of the most important transformative processes in the Earth system: coevolution. In the simplest terms, coevolution means evolving *together.* It is the reciprocal evolution of two or more entities, all influencing one another. Flowers and pollinators are a classic example, mentioned by Charles Darwin in *On the Origin of Species.* For tens of millions of years, flowers and their pollinators have shaped each other's anatomy and behavior, each pushing the other to aesthetic and adaptive extremes. Many insects, birds, and mammals evolved wings, mouths, eyes, and brains better suited to the quest for floral sustenance. In parallel, flowers became bolder, brighter, more fragrant, and more intricately shaped. Some flowers pool their nectar at the end of tubes so long and narrow that only a single species of moth, with an equally long tongue, can reach it. Mirror orchids trick male wasps into attempting copulation with them by mimicking the shape, color, scent, body hair, and even the shimmering wings of a female wasp; when the male wasp vigorously humps what he thinks is a mate, sticky sacs of pollen cling to his head. Other examples of coevolution include predators and prey as well as hosts and parasites.

Although coevolution usually refers to interacting species, it can also occur between other entities. Memes, technologies, and cultures can coevolve, for instance. Life and its environment evolve together, too. Darwinian evolution by natural selection happens through changes to the genetic composition of populations whose members vary in their traits. Those individuals best able to survive and reproduce in their particular environment leave behind the most offspring and pass on the genes coding for the very traits that made them so successful. Generation by generation, those genes and traits become more common in the overall population. Thus, species adapt to their environments. But their physical environments do not remain fixed during this process, nor are they subject to purely geological change. As living creatures evolve, they alter their surroundings extensively. Some of those changes persist and inevitably influence any evolution that follows. In this way, life becomes an agent in its own evolution. Although the enduring changes organisms make to their environments are not themselves genetically encoded, they are nonetheless passed from one generation to the next, becoming an important part of long-term coevolutionary processes. Natural selection is embedded within, and influenced by, the reciprocal transformation of organisms and their domains. Life and environment continuously shape one another and Earth as a whole.

In some respects, Sergey is the heir of Russian mineralogist Vladimir Vernadsky, who is himself a key precursor of James Lovelock. Although he remains obscure in the West, Vernadsky is revered in his home country, where his likeness appears on national stamps and memorial coins. There is a giant statue of him in Kyiv. An avenue in Moscow bears his name, as does a mineral (vernadite), several mountain peaks, a volcano, a lunar crater, and a species of algae.

Vernadsky was one of the first scientists to recognize life as a major geological force on our planet, an insight he developed in his 1926 book *The Biosphere*. Geologist Eduard Suess coined the term *biosphere* in a passing remark in 1875, but he never formally defined the concept or expanded his thinking. Vernadsky envisioned the biosphere as a fundamental layer of the planet containing life—an envelope extending from Earth's crust to the edge of the atmosphere. Within the biosphere, life dramatically altered the flow of energy and matter. "An organism is involved with the environment to which it [has] not only adapted, but which is adapted to it as well," he once wrote. "Since life on Earth is viewed as an accidental phenomenon, current scientific thought fails to appreciate the influence of life at every step in terrestrial processes. . . . As traditionally practiced, geology loses sight of the idea that the Earth's structure is a harmonious integration of parts that must be studied as an indivisible mechanism."

By the time Sergey had put together many of the pieces of his theory, it was the late 1980s and he was running the Northeast Science Station in Chersky. Nikita was still a toddler. When Sergey looked at the land surrounding the station, he saw a frozen desert. Gangly larch trees, woody shrubs, and rootless carpets of moss covered the permafrost. Compared to grasslands, this was a sluggish and nutrient-poor ecosystem. In the Ice Age mammoth steppe, herbivores would have swallowed plants by the chompful and broken them down in the hot, wet, microbial vats of their multi-stomached digestive systems. Carbon, nitrogen, potassium, and other essential elements would have cycled swiftly from plant to animal to air and soil and back again. In the boreal forests of modern Siberia, larch needles and waxy shrub leaves remained on the ground for decades, slowly decomposing. Apart from vexing clouds of mosquitoes, there was little wildlife to be seen. Most of the Arctic had essentially become "weeds covering the cemetery of mammoth steppe," Sergey told me.

He noticed, however, that wherever fire or human activity disturbed the soil and displaced the moss, grass flourished. And then there was Wrangel Island, with its pastures, herds of muskoxen and reindeer, and flocks of birds—all of which indicated that the Arctic could still support a large and diverse population of animals. What if, Sergey wondered, it were possible to re-create the mammoth steppe? Grazers had maintained their Pleistocene ecosystem for hundreds of thousands of years. If he brought herbivores back to the Arctic, maybe they could do so again. It would be a perfect way to test his theory.

Sergey proposed his idea to a few colleagues, who helped him present it to some of the leading Russian scientists at the time. Impressed, they agreed to a small field experiment. Within weeks, a paddock was constructed near the research station in Chersky and helicopters flew in twenty-five Yakutian horses, a large Siberian breed that can survive without shelter in temperatures as low as −94°F and is content to eat frostbitten turf. In a matter of months, the horses had trampled most of the moss and destroyed many of the shrubs within their enclosure. Grass began to grow. Levels of nitrogen and phosphorus increased tenfold. Then, in 1991, the Soviet Union collapsed and the government support for Sergey's newest study evaporated.

Sergey persisted. He knew that in order to continue he would have to secure funding from wherever possible and persuade the international scientific community that his research had merit. It occurred to him that his experiment could do much more than restore a lost ecosystem. By the 1990s it had become obvious to Sergey that the permafrost in the Arctic was thawing due to global warming. Some parts of the Arctic had already morphed into fetid marshland. Sergey realized that if enough permafrost thawed, enormous quantities of carbon dioxide and methane would likely seep into the atmosphere, potentially triggering runaway warming and a new period of climate cacophony.* The resurrection of the mammoth steppe could thwart this awful fate by refreezing the permafrost and stabilizing the global climate, just as it had done in the Pleistocene.

In convincing other scientists of the connection between his research and climate change, Sergey saw a chance to give his project renewed urgency and importance. At the time, he was one of the few researchers who recognized the immense danger posed by thawing permafrost. No one had the raw data to prove it. So Sergey spent the next seven years measuring levels of carbon and methane stored in frozen Arctic soils and tracking their dissolution. He demonstrated that

* Some scientists have since challenged this idea. Exactly how much carbon dioxide and methane would be released by melting permafrost, the speed with which it would escape, and the extent to which it would exacerbate global warming are matters of ongoing investigation and debate.

Arctic permafrost contained at least one trillion tons of carbon, twice previous estimates and more than all of the carbon in fossil fuel emissions since 1850 combined. As the uppermost layers of Arctic soil warmed and the permafrost below them thawed, microbes roused, multiplied, and began eating the vast stores of organic material, producing carbon dioxide, methane, and heat in the process. In a self-amplifying feedback loop, the heat generated by microbial activity accelerated the thawing of permafrost, which stimulated even more microbial consumption. But if heat-trapping snow were removed—or compacted by a mere four inches—the permafrost would cool by 1.8 to 3.6 degrees Fahrenheit, enough to keep it frozen.

Sergey's pioneering and meticulous studies brought him new levels of fame and respect. In the late 1990s, the Russian government gave him a little over thirty-five thousand acres of protected tundra and boreal forest surrounding the Chersky research station to use for his experiment. He formally designated the area Pleistocene Park, a winking reference to a certain science fiction franchise based on the premise of reviving dinosaurs. By 1998, Sergey had a detailed vision of his ambitious project and more than enough land to get started. Now he just needed some animals.

NIKITA ZIMOV WAS RUNNING across the chapped terrain of Wrangel Island, wrapped in curtains of fog, when he thought he heard a polar bear. Dread pressed in on him from all sides. Silently, he cursed his foolishness.

After losing the baby muskoxen, the Zimovs, their companions, and Gruzdev had boarded ATVs and begun a search. At some point, Nikita, proud of his youth and fitness, had decided to run alongside the vehicles. Now, he was lost, alone, and, in his own typically understated words, "a bit worried." When he had agreed to help his father with Pleistocene Park after graduating from college, this was not what he had envisioned. He tried to remember everything he knew about how to survive an encounter with a polar bear. He recalled someone telling him that the most effective polar bear deterrent was a fire extinguisher, because it produced a sound similar to the hissing noise that rival male bears made to convey their size and strength. But he didn't have a fire extinguisher. He didn't have any way to protect himself or anywhere to hide. So he kept running through the fog until, by chance, he found the nearest ATV and hastily climbed aboard.

Whenever the Zimovs and their search party located a herd of muskoxen, they approached it slowly and cautiously. Threatened muskoxen move with martial precision: the adults form a protective ring around the calves, keeping their gaze fixed on the danger and shifting position as a unit when necessary. Sometimes the largest bull will orbit the herd, ready to charge as a last resort. His curling, sharp horns can easily puncture any assailant.

The men would enclose the oxen's circle with one of their own, surrounding the herd on ATVs so that Gruzdev could shoot a tranquilizer dart at one of the calves. A few minutes later, after the calf collapsed, the search party would slowly push the herd away. One of them, often Nikita, would lay his weight against the calf, hands and face buried in its plush brown fur, waiting for it to wake up. Once the calf regained consciousness, which usually took several hours, Nikita and the others would grab it by the fur and steer it into a trailer hitched to one of the ATVs so that they could bring it back to the mended—and now guarded—paddock. Catching even a single animal in this manner was extremely challenging. Fog often reduced their visibility, and rain drenched their clothes. The tranquilizers were sometimes too weak to render a calf fully unconscious. And it could take five or six hours just to find a herd. After eight days of roaming the tundra, they finally recaptured six baby muskoxen.

In mid-September 2010, the Zimovs corralled the muskoxen into crates, loaded them onto their boat, and left Wrangel Island. The trip back home was not as tranquil as their outbound journey. In the midst of a storm, their electronic devices became unusually temperamental. Batteries were running low. The GPS began to malfunction, sometimes consenting to give them coordinates but failing to indicate in which direction they were traveling. To compensate, Nikita fashioned a makeshift weather vane by tying a strip of cloth to a fishing pole at the front of the boat. In an attempt to conserve battery power, he kept the GPS switched off, turning it on only every hour or so to quickly note their latest position on the map. Maneuvering at night was especially risky. Without much light or proper navigational tools, they were essentially boating blind.

Although the waters they traversed were ice-free, an intense storm developed on the second day of their voyage. Their refurbished Soviet boat was forced to ride huge swells—nine feet high at least. As they slid down the back of one monstrous wave after another, the little plastic boat they were towing behind them—a backup in case of emergencies—slammed into the stern of their main vessel, swinging and bouncing wildly. Everyone onboard was seasick for days, including the musk-oxen, which lay flat and noiseless on the deck, not even stirring for their oats and hay. A day later, the sea lulled, the sun emerged, and the Siberian mainland surfaced on the horizon. They knew it would be only a few more days before they reached Chersky. All they had to do was follow the coastline.

THE THEORY THAT SERGEY developed in the 1970s regarding Pleistocene megafauna prefigured an emerging consensus in ecology. Scientists have long recognized that plants reshape the planet's land surfaces. After all, plants dominate not only the continents but also the biosphere as a whole. There are about 550 gigatons of carbon-based biomass on Earth, of which plants comprise 450 gigatons—over 80 percent. In contrast, animals make up less than one half of 1 percent of Earth's biomass, concentrated largely in the ocean in the form of fish and invertebrates. Perhaps in part because of these large discrepancies, ecology has historically overlooked and undervalued the ways in which animals shape the planet's landmasses.

Certain exceptions were recognized, however. Charles Darwin was one of the earliest scientists to seriously consider the possibility that animals could change the planet's topography. His principal example? Earthworms. By constantly tunneling through soil in their native ecosystems—digesting huge quantities of dirt, decomposing organic matter, secreting slime, and depositing water-retentive castings—earthworms improve soil's granular structure, mix its different layers together, and open channels through which oxygen, water, and nutrients can flow.* Darwin described the earthworm as the "unsung creature which, in its untold millions, transformed the land as the coral polyps did the tropical sea." Not long before his death, he wrote an entire book about earthworms, which became a surprise bestseller.† Some of his scientific peers were skeptical of these claims, however; others dismissed them entirely. Yes, they conceded, certain animals could alter their environments, but only in small, local, and rather obvious ways. For more than a century, Western science minimized the importance of earthworms and other animal geoengineers, regarding them as asterisks to the prevailing geological doctrine, which emphasized the role of inanimate forces.

In recent decades, however, scientists have made remarkable discoveries about the many ways that animals small and large rework the planet's land surfaces, often with enduring consequences. Over the past half billion years, animals became the great movers and shakers of earth, continually redefining the planet's contours and thereby changing local ecology, global climate, and even Earth's evolutionary trajectory.

During the Pleistocene, giant ground sloths and armadillos—some larger than a modern elephant—likely dug branching burrows up to two thousand feet long. Researchers have uncovered hundreds of such paleoburrows in Brazil, complete with scores of claw marks on their walls and ceilings. In present-day South America, some species of leafcutter ant construct underground nests that span thousands of square feet and extend as much as twenty-six feet deep, requiring them to move more

* In contrast, certain introduced species of earthworm can severely disturb forest ecosystems with which they did not coevolve, in part by decomposing leaf litter too rapidly, depriving plants and other organisms of essential nutrients.

† This was *The Formation of Vegetable Mould Through the Action of Worms, with Observations on their Habits,* published in the fall of 1881, shortly before Darwin's death in the spring of the following year. "My whole soul is absorbed with worms just at present!" he proclaimed while composing the book.

than forty tons of soil. Soils that are replete with ants, termites, and burrowing rodents tend to be more stable, better drained, and more likely to retain nutrients. In North America, herds of migrating bison propel waves of springtime rejuvenation across the plains by intensively grazing and fertilizing grass, thereby encouraging the plants to continually produce palatable and nutritious young shoots. Collectively, bison exert a stronger influence on seasonal plant growth than weather or other environmental factors. Though there are fewer than eight thousand wild bison with the freedom to migrate across public lands in North America today, consider how powerful a force these animals must have been when there were still 30 to 60 million of them roaming the plains.

Beavers are perhaps the best-known ecosystem engineers. Many people are familiar with beavers' habit of felling trees and damming streams, which submerges the surrounding landscape in networks of ponds and canals that become vital habitat for numerous other species. Yet the full extent of beavers' topographical renovations is often underappreciated. Beavers have been reshaping the planet's surface by building dams for at least eight million years—and for most of that period, they were far more numerous than they are today. Beaver dams can stretch more than half a mile, exceed six feet in height, and last for centuries. As Ben Goldfarb writes in his eloquent paean *Eager,* "Beavers are nothing less than continent-scale forces of nature in large part responsible for sculpting the land upon which we Americans built our towns and raised our food. Beavers shaped North America's ecosystems, its human history, its geology. They whittled our world." A testament to this legacy is the crucial role beavers are playing in the ongoing restoration of Yellowstone's riparian ecosystems—a transition too often credited solely to wolves. A long history of wolf extermination allowed populations of elk and other herbivores to surge, which drastically reduced the abundance of riverside vegetation, such as willow, aspen, and cottonwood. Without the stabilizing roots of those water-loving plants, riverbanks collapsed and soil eroded. The reintroduction of wolves in the 1990s curtailed the elk population and encouraged willow to regrow, which in turn nourished beavers. In parallel, a massive beaver relocation project returned the semiaquatic rodents to Yellowstone's borders. As wolves and beavers repopulated the park, their combined influence rehydrated and revitalized many dry and degraded valleys.

Analogous chains of zoological transformation lace the atmosphere and ocean. Whales continuously travel between the ocean's dim depths and the sunlit surface, where they release what are scientifically termed fecal plumes (also known as poonamis), fertilizing photosynthetic plankton that are a vital component of the planet's carbon cycle. Whales also directly transport huge amounts of carbon to the deep sea. When a whale dies in the open ocean and sinks to the seafloor, it becomes an underwater oasis, its ample flesh and bones sustaining a profusion of peculiar worms, eels, crabs, and octopuses that never leave the abyss. Even the daily movements of whales—their gliding, diving, and breaching—maintain a more even distribution of nutrients in seawater than would exist without such agitation. By some estimates, the collective motion of ocean creatures mixes seawater as much as the wind and tides. Just above the waves, flocks of migratory seabirds drop huge quantities of nitrogen-rich guano on the cliffs and islands where they nest—an important nutritional link between land and sea. In the Arctic, bacteria break down the guano, which releases ammonia, which combines with other compounds in the atmosphere to produce tiny particulates that seed clouds. The resulting clouds reflect light and heat. So seabirds help keep the Arctic cool.

In the traditional framework of geology, rivers carry mineral nutrients from crumbled rocks to the sea, where ocean creatures consume them. When some of those creatures die and sink into seafloor sediments, they are subsumed by tectonic processes that melt and recycle rock, eventually returning the nutrients they contain to the planet's surface. "We are left with an impression that nutrient cycling in adjacent landscapes or gyres is disconnected except through the atmosphere or hydrosphere, and that animals play only a passive role as consumers of nutrients," write Chris Doughty, an ecologist at Northern Arizona University, and his colleagues in one study. As he and other scientists have discovered, that old portrait is greatly oversimplified. Animals play an especially important and unique role in the planet's nutrient cycles. Although animals are not nearly as abundant as plants in terms of biomass, as a group they are far more mobile and dynamic.*

Doughty and other ecologists are forming a new picture of the planet in which animals help nutrients flow from the ocean depths to continental interiors. Whales, jellyfish, and other marine creatures move nutrients to the sea's surface, feeding plankton, which in turn sustain fish and seabirds. Migrating seabirds, along with fish that swim up rivers to spawn, bring nutrients back to the continents. Bears, otters, and eagles eat spawning fish and drag their carcasses inland, where they decompose and nourish forests. Burrowing animals within forests and other terrestrial ecosystems improve soil conditions, benefiting plants. Plants crumble the crust into its mineral components, hastening the return of nutrients to the sea. Without such ecological loops, nutrients would be much more likely to settle in place, restricting life to small and isolated pockets of the planet.

In part by expanding and accelerating nutrient cycles, the development of large, highly mobile animals ultimately made Earth more habitable and resilient. "It now appears that the increase in Earth's ability to support life across time and, especially, its increasing ability to support complex, multicellular life are largely a consequence of biological processes," Stanford University Earth scientist Jonathan Payne and several colleagues have written. "These processes include natural selection for organisms with greater ability to survive in the face of environmental change but, more importantly, for ecosystems with greater complexity and stability as well as organisms that conduct activities that strengthen stabilizing feedbacks within the Earth system."

Animals have been changing the structure and chemistry of Earth's crust since their evolutionary debut. Six hundred million years ago, microbial mats likely covered much of the ocean floor, punctuated by

* As time-lapse videos reveal, plants move a great deal, too, albeit usually too slowly for us to see with the naked eye.

sessile, fernlike organisms that swayed in the currents. Other creatures resembling armored slugs and rippled pancakes—which are so bizarre that no one is sure whether to classify them as animals or something else altogether—slid along these primordial meadows, grazing as they went. Beneath the mats was a firm, oxygen-deprived substrate devoid of any life but bacteria. About 540 million years ago, during a burst of evolutionary innovation called the Cambrian explosion, very different creatures emerged: burrowing worms, beetle-like trilobites, giant shrimp with toothed tentacles, and a mind-boggling creature, aptly named *Hallucigenia,* that a modern observer might describe as the result of a hot dog's unfortunate encounter with a porcupine.

In a geological blink, most of the strange organisms that evolved before the Cambrian explosion went extinct. Some paleontologists have argued that the new, more athletic Cambrian animals outcompeted their Ediacaran predecessors, which were viewed as "failed evolutionary experiments." Recent evidence suggests an alternative explanation: Cambrian creatures caused a mass extinction by dramatically reengineering their physical environment. Increasing predation during the Cambrian Period spurred the evolution of armor, such as bristles, spines, and shells. These novel mineralized appendages allowed animals to perturb the ocean floor much more effectively than ever before, burying themselves to hide or digging through it for food. All those early worms and arthropods tore apart the microbial mats, kicked up huge amounts of sediment—which may have clogged the filtration systems of sessile creatures—and irrigated the seafloor with tubes and tunnels that scientists have likened to "a system of veins and arteries."

Microbial mats had sealed much of the ocean bottom for eons, but now oxygen and nutrients flowed more freely through the sediment, allowing life to permeate and occupy its many layers. As novel groups of organisms adapted to the newly turbulent seafloor, diversifying into new species, the mat-loving creatures died out. Thanks in part to this major ecological shift, known as the Cambrian substrate revolution, the modern ocean is far more habitable and biologically diverse. Although microbial mats still exist, they primarily inhabit extreme environments, such as hypersaline lagoons and oxygen-poor basins where sediment-disturbing animals cannot live.

Within the span of recorded history, the capacity of animals to shape the planet's land surfaces has been particularly conspicuous in Africa, which is home to the largest living terrestrial species. In the late 1880s, while trying to conquer Ethiopia, Italy supplied its armies with Indian cattle for food and labor. Arriving at the port of Massawa, the cattle brought a highly contagious viral disease known as rinderpest (German for "cattle plague"), which spread rapidly throughout eastern and southern Africa. Rinderpest killed more than 90 percent of domestic cattle and huge numbers of wild herbivores, ultimately starving one third of Ethiopians and two thirds of the Maasai in Tanzania to death. The number of wildebeest, one of the Serengeti's most important grazers, plummeted from more than a million to about a quarter of a million. Without the usual pressure from herbivores, grass and shrubs grew out of control, fueling larger and more frequent wildfires. Up to 80 percent of the Serengeti was incinerated each year, adding large volumes of carbon dioxide to the atmosphere. Young trees began to burn down before they could grow tall enough to escape the fields of flame. The Serengeti had long been a mosaic of savanna and woodland. By 1980, many historically wooded areas were treeless.

Around the same time, however, wildebeest populations were recovering, due to campaigns to vaccinate domestic cattle against rinderpest, which prevented transmission of the disease to wild herbivores. As the wildebeest returned, the grass receded to its usual levels. In turn, wildfires became smaller and more intermittent, which allowed trees to grow tall enough to survive future blazes. Gradually, woodlands reclaimed their ancestral territories. Today, the Serengeti is once again a carbon sink, meaning it absorbs more carbon from the atmosphere than it releases; in doing so, it offsets all of east Africa's annual fossil fuel consumption.

Elephants, too, modify Africa's savannas and woodlands: they eat huge quantities of vegetation, fell trees, dig waterholes with their tusks, and scatter seeds in their copious dung. Given their size, many of these changes are hard to miss. Recently, however, researchers have discovered that the mere impression of an elephant can alter the landscapes they inhabit and change the fate of other species. In 2014, while on a field course in Kibale Forest National Park in Uganda, a young biologist named Wolfram Remmers noticed some dragonflies hovering over a puddle of groundwater that had formed in an elephant's footprint. It was far from alone. Other areas of the forest were also indented with hollows made by the thick round pads of elephant feet. Each hollow contained as much as fifty gallons of water.

Remmers wondered what might be living in and around these miniature ponds, so he acquired a kitchen sieve and began to explore. The puddles had essentially become ecosystems in their own right, sustaining multitudes of microorganisms, beetles, mites, mayflies, worms, leeches, snails, and dragonfly larvae. In some regions of the forest, the water-filled footprints—which can persist for a year or longer—were the only ponds available to such creatures. Similar research has found that the tracks of Asian elephants are a critical habitat for frogs, especially during the dry season.

The footprints of elephants, mammoths, and other multi-ton megafauna have almost certainly served as impromptu ecosystems for tens of millions of years, yet prior to a decade ago, few scientists had formally documented them. Life's influence on its environment is so thorough and manifold that we are still discovering all the forms it takes, even when it comes to the largest and most dynamic creatures among us. Simply by taking a step, an animal can remake the earth and leave new worlds in its wake.

TODAY, PLEISTOCENE PARK ENCOMPASSES about five thousand fenced acres inhabited by more than one hundred herbivores: horses, reindeer, elk, sheep, yaks, cows, and bison. In the outer rings of the park there's at least one wolverine, some Arctic foxes, and several brown bears. Those who have heard of Pleistocene Park usually remember one thing above all else: that it will be the future home for resurrected mammoths. A few scientists are genuinely interested in using genetic engineering to bring back the mammoth—part of a larger movement known as de-extinction—but reanimating Ice Age creatures has never been the Zimovs' goal. "All the media talk about mammoths to attract readers and viewers," Nikita has said. "Our work started before people even started thinking about doing any mammoth cloning research. . . . If someone should knock on my door in the future and say that he brought me a mammoth, I would be happy to take this mammoth to the park. Pleistocene Park would probably benefit from it. But our work is independent, and we can achieve our goals without mammoths too."

Some of their true ambitions have already been realized. On the second day of my visit, Nikita and I traveled an hour by speedboat along the Kolyma River from the research station to the park itself. As soon as we docked, two large, muddy dogs bounded toward us in greeting, companions of the rangers that live onsite to look after the wildlife. The unmarked entrance to the park was framed by a house built atop shipping crates; nearby were several shacks, some rusty boats, and a wood and wire pen that at the time was used to hold sick or injured animals. Nikita and I began hiking into a dense patch of willow shrubs and pole-thin larch trees with peeling gray bark. The ground was heaped with piles of dry pine needles that seemed to resist decay. "Not a single animal eats these plants," Nikita said. "There's not much going on in this ecosystem."

We walked a few yards away to a markedly different landscape streaming with lush grass and ornamented with pink and yellow wildflowers. Here and there, I noticed some lingering willow with shorn branches and leaves. "This was the same forest five years ago," Nikita said. "Identical." Grazing by cows, yaks, and sheep along with grass seed spread by hand had transformed the forest into grassland. Larch forest doesn't sequester a lot of carbon, Nikita explained, because the trees are so small and their roots so shallow. Even a larch tree two centuries old might have a trunk not much wider than a table leg, a fraction of the size many other tree species achieve by that age. "The albedo, you can actually see," he continued, gesturing to the stark contrast between the dark forest and pale grassland in the distance. Nikita and his colleagues have found that, depending on the season, the upper layers of grazed soils are as much as 23.4°F colder than ungrazed soils and 3.8°F colder on average. Grazed soils also store more carbon due to enhanced fertility and root growth.

Yet the gap between Pleistocene Park and the mammoth steppe is undeniably immense. Scientists estimate that during the Pleistocene at least one billion individual megafauna roamed the continents, a large portion of which lived in northerly grasslands. When Sergey talks numbers, he does so with a nonchalance that belies the daunting scale of his mission. He says he would be extremely pleased if modern Siberia were populated with a mere 50 million large herbivores, which he thinks would be enough to keep all of its permafrost intact. For now, he would be content to have one million animals in his park, which he says would be sufficient for a stable ecosystem. But he hasn't reached even one tenth of one percent of that goal.

The Arctic is the ideal place to conduct an experiment of this magnitude: there is so much land, the vast majority of which is wilderness. Yet its size and remoteness make the endeavor all the more challenging. Even if the Zimovs suddenly receive an extremely generous grant or an ark of donated animals, there is still the small matter of transporting them safely to Siberia. For more than two decades, the Zimovs have funded their project primarily with their own money and a few thousand dollars earned from crowdsourcing campaigns. They acquired many of the hundred or so animals in the park today through perilous and self-organized expeditions, like the nearly disastrous voyage to Wrangel Island. Sergey is well respected in the scientific community as an expert on Arctic ecology—and Pleistocene Park has its share of ardent fans—but most of his peers regard his grand ambitions with a kind of leery admiration. Even if they applaud the intentions and the science behind them, they simply do not think the overall scheme is feasible.

"Some people do not believe the park is possible to do practically," Nikita says. "'It's too much effort. Climate change is already here. You will not make it in time'—and so on and so forth. Maybe they are right. The thing is, if I don't do anything, nothing will happen. Unless we are doing something—not just writing about it, not just shouting or yelling that 'Oh, we're all going to die,' but actually, practically doing something—nothing will happen."

One of the Zimovs' keenest desires and greatest challenges has been the introduction of bison to the park. Bison, which were abundant on the Pleistocene steppe, are some of the largest extant mammals that can tolerate a Siberian climate. A few months before my arrival, Nikita purchased twelve young bison from a farmer in Denmark and transported them to Chersky by truck and river barge, a grueling journey that took five weeks. As soon as he released the bison, they ran straight into a lake, not realizing what it was, then lumbered back onto land, hiding in the shrublands. Although traces of the bison were abundant throughout the park, the Zimovs had not encountered the animals themselves since their arrival and wondered how they were adjusting.

On one of our visits to the park, Nikita, Sergey, several rangers, and I commenced a search for the bison. The frequent use of ATVs had opened a corridor through the forest, which we followed, sidestepping half-dried tracts of mud and pools of putrid water edged with star moss and reeds. The air was livid with mosquitoes, which swarmed us relentlessly, undeterred by nets and long sleeves. Though we found telltale footprints, droppings, and twigs wrapped in strips of brown fur, we neither saw nor heard the bison.

Eventually we reached an undulating field with a distinctly Seussian appearance, filled with large tufts of grasses known as tussocks, each sprouting from a clod of dirt shaped like a giant mushroom stem. Traversing the field proved a gymnastic endeavor, as it was so easy to topple a tussock or sink into the hidden gaps between them. Once across, Nikita launched a drone he had brought with him, using its camera to search for our quarry—something he had tried several times earlier with no success. This time, however, he found the bison just a short distance from where we stood.

We moved quickly, splitting into two groups in order to herd the bison toward a paddock, where it would be easier to check on their health. Nikita and the two rangers headed left, into the thick of the forest. Sergey and I veered to the right, remaining in a largely treeless alley near one of the park's fenced borders. When I asked the elder Zimov what I should do, he gave me instructions in a hushed voice: "Follow me. Stay one meter behind. If you see bison, don't move absolutely. If I stop, you stop also. If you afraid something, jump on the posts for the net."

We moved forward cautiously, noting piles of fresh dung. Sergey stooped to examine a fallen tree branch and, seemingly satisfied, carried it with him. Suddenly Nikita and the rangers yelled in alarm. They had spotted the bison. "Haup! Haup!" they cried, whooping and whistling to alert the animals to their presence and keep them moving. A frenzy of bison crashed through the forest, halting directly in front of Sergey and me. About ten in number, they had petite horns, wary eyes, and short-haired coats as dark as fresh basalt. Though young, they were formidable, especially as a group. They shook their heads and stamped the ground. It seemed they might charge us at any moment.

Sergey held out the tree branch, Moses-like, and calmly addressed the creatures before us. "Little bison, little bison," he said in Russian. "Don't come this way." They retreated to the forest, where Nikita and the rangers prevented them from fleeing further. Every time the bison emerged from the trees, Sergey stood his ground. Every time the bison tried to run in the other direction, Nikita and the rangers blocked their path. In this way, we gradually pushed the herd forward and finally into an unused paddock. The Zimovs and rangers swiftly sealed the only opening with a barrier improvised from wood and wire. Restless, the bison rushed from one corner of their enclosure to another, occasionally throwing their weight against it. Although segments of the paddock were nothing more than stacked branches, it held, and the bison eventually settled down.

As the rangers tended to the animals, Sergey and I continued to explore the park, walking through a typical tract of willow and larch, across a river, and into 120 acres of rolling grassland. In the past, Sergey explained, this area had been a bog. Grazing animals had promoted the growth of grass, which had increased the rate of transpiration and siphoned away the excess water. It was now the largest unbroken expanse of grass in the park. "Here," Sergey said, spreading his arms, "is the future of this landscape."

We strolled through the rippling grass toward the park's entrance. In the distance we saw a herd of silky brown cows, sheep the color of lightly roasted marshmallow, and a Yakutian horse so fair and sinewy it might have been a marble carving come to life. Unlike the bison, these veteran residents of the park were unperturbed by our presence. They seemed entirely comfortable in this refuge at the edge of the world, as much a part of it as the sky and soil and the grass they had helped cultivate. They had become the stewards of their realm—the architects of their own Eden.

As we approached, I was particularly drawn to a yak with a coat of mottled cream and cinnamon. Her curtains of fur were so gloriously long and thick that they obscured her belly and flowed halfway down her face. I moved closer until I was just a few feet from where she was grazing. She didn't bellow or startle or move a single limb. For several minutes she didn't seem to notice me at all. Eventually she lifted her head, tossed her bangs to the side, and appraised me with an obsidian eye. Then she lowered her gaze and resumed chewing.

CHAPTER

3

A GARDEN IN THE VOID

MY FIRST GARDEN WAS A FOUR-BY-SIX-FOOT RECTANGLE OF FALLOW earth on the side of my family's house in California, which I adopted at the tender age of twelve. I dug out all the weeds, tilled the soil, and grew various herbs and vegetables from seed, including chives, parsley, radishes, and tomatoes. I remember the particular thrill of watching corn seedlings erupt into robust stalks twice my height and the satisfaction of plunging a spade into the ground and lifting out carrots nearly as thick as those at the supermarket. I relished the experience so much that I wrote an embarrassingly earnest, but mercifully brief, article for the opinion section of the *San Jose Mercury News* in which I encouraged readers to embrace gardening and reap "the veggies" of their labor (my mom still has the clip).

For close to a decade after leaving home for college, I rarely had the chance to garden. When I moved to Cambridge, Massachusetts, for my first reporting job, I applied for a plot in a community garden. I received only one reply, three years later, informing me that there were still no spaces available. Later, while renting a ground-floor apartment in Brooklyn, New York, I briefly helped tend a yard bordered by strips of hostas, lilies, and hydrangeas. It was not until my early thirties, however—several years after I had moved back to the West Coast—that I was finally able to cultivate a garden on my own property.

In the summer of 2020, my partner, Ryan, and I bought a house in Portland, Oregon. We were especially excited about the spacious, south-facing backyard, which at the time consisted of nothing more than a tool shed and a derelict lawn. For us, it was an ideal canvas on which to create a garden from the ground up and an opportunity to change our personal relationship with the planet. Maybe we weren't visionary scientists attempting to transform the Arctic landscape. Perhaps we didn't have access to supercomputers that could model global climate, let alone the expertise to use them. But we did have this plot of land—this small piece of our living Earth—and the freedom to help it flourish. Where once there was nothing but crisped turf, we could establish a biodiverse, carbon-storing wildlife habitat adapted to a rapidly shifting climate.

We began planning immediately. Ryan, who had studied art and architecture in his youth, drafted several versions of our design from different perspectives. To give the long, rectangular plot a more inviting geometry, we envisioned a winding flagstone path partly screened by small trees and shrubs. Near the front of the garden, we would form a pond, which would be counterbalanced by a rock garden on the other side of the path. Toward the back, where there would be nothing to obstruct incoming sunlight, we'd build a series of raised beds for herbs, berries, and vegetables. We'd train espalier fruit trees along the fence. And all throughout the garden, we would plant long-blooming perennials beloved by pollinators.

One of the first and most essential steps was removing the existing lawn. Unlike long grasses and sedges that are allowed to sprawl, flower, and set seed, cropped turfgrass offers little food or habitat for wildlife while simultaneously guzzling water and fertilizer. Western culture tends to regard a verdant lawn as a fecund space—a symbol of wealth and vitality—yet it is often the most sterile and impoverished part of any garden.

In late summer, we hired a landscape contractor, Ted, to remove all two thousand square feet of turfgrass from our backyard. While waiting for that work to begin, I decided to tackle the lawn in our much smaller front yard myself with a pickaxe and shovel. We intended to plant a woodland shade garden beneath a Douglas fir on the western side of the front yard and a bed of colorful, sun-loving perennials in the brighter eastern half.

As soon as I began to slice and lift the sod, exposing the ground beneath, I started to worry that our grand plans for a resplendent garden were naïve and futile. I did not know much about soil at the time, but prior experience had taught me that ideal garden soil was soft, dark, and crumbly. In contrast, the soil I'd uncovered was dry, sallow, and compacted. A spade was useless. Even with a pickaxe, I struggled to dig deeper than a few inches, frequently striking rocks, bricks, and chunks of cement. When I squeezed lumps of soil in my hand, they felt as hard as granite. The few small clods I was able to crush exploded into dust.

The situation in the backyard was no better. Early one morning, during one of several heat waves that year, Ted maneuvered a Kubota compact excavator through our garden gate and began using its metal claw to rip the lawn from the ground. The soil his excavations revealed was similar to what I had exposed out front: hard, dry, and full of rubble. Ted attempted to rake and loosen the bulk of the soil with his machinery, but some areas were so unyielding that he nearly gave up in frustration. "It feels like I'm trying to dig through concrete over here!" he said one day, wiping streams of sweat from his face and neck.

By speaking with our neighbors and studying archived images on Google Maps, we learned that our lot and the one directly east of us had once been a single property with not much in the way of a garden beyond a haggard lawn, part of which had served as a parking lot. For several years, as developers remodeled our neighbors' house and then built the one we bought, our backyard had been a construction site. The new lawn on our property was a facade—a thin green frock disguising more than a decade of neglect.

After several days of excavation, I walked into the middle of our future garden, rested the pointed tip of a shovel against the ground, and pushed all my weight against it. It barely budged. I repeatedly stomped the shovel as hard as I could, finally sinking it a couple inches below the surface. Kneeling, I scooped up a layer of coarse, dull soil and brought it within inches of my face, as though it would divulge all its secrets if I just stared long enough. At this point, I was approaching the threshold of despair. Clearly Ryan and I had been too hasty in our horticultural ambitions. We'd invested so much time and energy in planning the shape and feel of the garden, but we'd failed to properly consider its foundation. I scratched at the earth with my bare fingers, searching for an ant, a worm, a root—for any trace of life. Nothing.

What could possibly grow here?

THE HISTORY OF DEGRADATION that we uncovered on our property is a microcosm of what our species has been doing to the planet's land surfaces for millennia. Like so many animals before us—from termites to four-ton ground sloths—humans have radically altered Earth's crust and soils. The specter of environmental devastation is often accompanied by images of smoke-spewing factories and concrete megalopolises, yet cities, roads, railways, mines, power plants, and other human infrastructure occupy less than 3 percent of the planet's habitable land. The vast majority of acres modified by human hands are devoted not to dwellings or energy production but to agriculture—to gardening writ large. A thousand years ago, humans used less than 6 percent of Earth's ice-free, non-barren land for farming. Today, about half of the planet's habitable land is used to grow crops or raise livestock.

The first farmers likely used digging sticks, hoes, and other simple tools made from wood and stone. As far back as 171,000 years ago, Neanderthals in what is now Italy used stone and fire to craft three-foot-long digging sticks from boxwood, which they may have used to forage roots and tubers, grind plants, and club small burrowing animals. Much later, though it's not clear precisely when, humans developed one of the most consequential pieces of technology in history: the plow. As geomorphologist David R. Montgomery has written, the invention of the plow not only revolutionized human civilization—it also "transformed Earth's surface."

Scratch plows, or ard plows, which emerged at least six thousand years ago in Mesopotamia, were constructed from wood and pulled by humans or animals, forming shallow grooves in the soil in which to plant seeds. Over time, plows became larger and stronger, incorporating stone and eventually metal. In India, which has a particularly long history of sophisticated metallurgy, people may have been making plows with iron shares—wedgelike blades that slice through the ground—2,700 years ago. Much later, the widespread adoption of the moldboard, a curved metal plate sitting atop the share, enabled plows to invert soil's upper layers, conveniently burying weeds and the remains of earlier crops.

With the advent of the plow, humanity began to confront one of agriculture's central dilemmas: repeatedly tilling a tract of land ultimately destroys its fertility. In the short term, tillage—agitating soil for the purpose of cultivation—provides numerous benefits to farmers: it loosens soil, suppresses weeds, incorporates manure and other amendments, and facilitates germination and early root growth. In the long term, however, tillage severely disrupts the soil ecosystem, depriving it of symbiotic plants, fungi, and microbes and increasing its susceptibility to erosion by wind and water. Continually clearing and tilling a plot of farmland is the equivalent of bulldozing a forest year after year, leaving the soil weak and unprotected. When soil is stripped of its vegetal armor, even a little bit of wind or rain can be catastrophic, shattering and dispersing the granules that give it structure. It was exactly that kind of vulnerability, combined with extreme drought, that culminated in the Dust Bowl of the 1930s.

The fundamental design of the moldboard plow endured for millennia, allowing people across the globe to cultivate previously unworkable earth. "The plough is to the farmer what the wand is to the sorcerer," Thomas Jefferson wrote in 1813. In the mid-nineteenth century, the Agricultural Revolution and the Industrial Revolution collided in the form of the first commercial steam-powered plows, which were soon superseded by tractors and other heavy machinery equipped with gas-powered internal combustion engines. Mechanized equipment allowed farmers to expand into even tougher soils and dramatically increased their overall efficiency, especially in wealthy industrial nations.

At the same time, fossil-fueled machines hastened the destruction of fertile soil. On average, agriculture, overgrazing, deforestation, and other forms of human land disturbance erode soil ten to thirty times faster than it is generated, removing centuries' worth of accumulated soil in less than a decade. A 2021 study revealed that about one third of agricultural land across the Corn Belt in the United States has already lost all of its topsoil. In some regions of Africa and Asia, soil is disappearing a hundred times faster than it can be replaced. Around the world, close to a third of conventionally farmed soils have a lifespan of less than two hundred years. Without adequate interventions, 16 percent of such soils will be gone within a century.

One of the most severe repercussions of widespread soil erosion is a rapid depletion of nitrogen, which is essential for plant growth. In many pre-twentieth-century societies, farmers relied on a small selection of potent fertilizers, such as fossilized dung, bird guano from islands off the coast of Peru, and saltpeter (sodium nitrate) mined from the Atacama Desert in Chile, which had up to thirty times the nitrogen of barnyard manure. By the late nineteenth century, these limited and obscure resources were thought to be on the brink of exhaustion, generating waves of alarm. European nations such as Germany and the United Kingdom were especially concerned as they already imported a large portion of their grains and were rapidly running out of arable land. In 1898, William Crookes, president of the British Association for the Advancement of Science, warned that the world's wheat-producing soil was "totally unequal to the strain put upon it" and that "all civilized nations stand in deadly peril of not having enough to eat." Crookes predicted a global deficit of wheat as early as 1930—unless someone discovered a new way to deliver nitrogen to crops.

All life requires nitrogen, a primary component of genes, proteins, and enzymes. Although nitrogen is abundant on Earth, comprising 78 percent of the atmosphere, it is largely inaccessible to most organisms in its gaseous form. Pairs of nitrogen atoms in the atmosphere are linked by one of the strongest molecular bonds in existence. Lightning is one of the few physical phenomena powerful enough to split that bond. Because gaseous nitrogen is so difficult to break apart and mix into new molecules, it is useless to most living creatures. Confronted with this daunting challenge, Earth evolved an elaborate set of interconnected processes through which its ample reservoir of nitrogen is continually converted from one chemical form to another and cycled between the animate and inanimate in air, sea, and land. Microbes are critical to this cycle: bacteria and other microbes are the only organisms that have evolved enzymes with the ability to cleave atmospheric nitrogen and turn it into biologically useful molecules, such as ammonia, nitrite, and nitrate. Some of these nitrogen-fixing microbes, as they're known, live symbiotically on the roots of peas, beans, and other legumes, whereas others live independently in soil and water. Microbes also decompose the nitrogen-filled remains of plants, animals, and fungi and return nitrogen to its gaseous state. All complex life depends on the chemical wizardry of these nitrogen-manipulating microorganisms.

In the early twentieth century, however, our species discovered a new way to split apart nitrogen gas and synthesize ammonia—an unprecedented event in the history of complex life that radically altered Earth's chemical cycles. In 1907, through similarly innovative methods, German chemists Walther Nernst and Fritz Haber independently demonstrated the use of intense heat and pressure to separate atoms of gaseous nitrogen and recombine them with hydrogen, thereby producing ammonia. Haber and Carl Bosch of the chemical company BASF, along with Bosch's assistant Alwin Mittasch, adapted the process for industrial-scale production, in part by introducing more suitable catalysts. By 1913, a plant in Oppau in southwestern Germany was generating 7,000 metric tons of ammonia each year. A few years later, an even larger plant in Leuna in eastern Germany achieved an annual production of 146,000 tons.

The Haber-Bosch process, as it became known, is now regarded as the most important industrial process ever developed. At first, Germany primarily used ammonia synthesis to produce more explosives, prolonging World War I.* After Bosch revealed the details of the pro-

* A staunch patriot, Haber also developed poisonous gases for use in World War I and supervised their deployment. In May 1915, his wife, the chemist Clara Immerwahr Haber, shot herself with an army revolver, possibly in part due to her objections to his work on chemical weapons.

cess during negotiations at Versailles, other countries began synthesizing ammonia as well. With some modifications, the Haber-Bosch process provided an entirely novel and highly reliable source of nitrogen fertilizer, averting the imminent global food crisis and supporting a massive expansion of the human population. During the twentieth century, the world's cumulative yields of staple cereals increased sevenfold and the total population swelled from 1.6 billion to 6 billion. An estimated 50 percent of the nitrogen in all human bodies across the planet now derives from the Haber-Bosch process. Without synthetic nitrogen fertilizers, today's global crop harvest would be halved and two out of every five people currently alive would not exist.

These historic shifts were not exclusively due to the synthesis of ammonia, however. In the mid-twentieth century, the Rockefeller and Ford foundations, among other organizations, funded plant breeding research that created much higher-yielding varieties of wheat, rice, maize, and other staple crops, some of which matured more quickly, produced multiple crops each year, or had shorter and sturdier stems, supporting more grains per head. The development and distribution of these higher-yielding crop varieties, coupled with the widespread adoption of synthetic fertilizers, pesticides, irrigation, and mechanized farm equipment, is known as the Green Revolution. These advances dramatically increased crop yields and curbed hunger and malnourishment in China, India, Brazil, Mexico, and much of the developing world, with the major exception of sub-Saharan Africa, where high transport costs, limited irrigation, a lack of infrastructure, and inequitable pricing policies hindered their success.

In addition to saving more than a billion people from famine, the Green Revolution spared much of the world's wilderness and soils from destruction. Historically, the only way to feed growing populations was to expand farmland. The Haber-Bosch process and the Green Revolution partially liberated humanity from these constraints. Farmers can now produce almost three times more cereal from a given tract of land than in the 1960s. Without these advances, at least 1.48 billion hectares of wilderness—an area the size of the United States and India combined—would have been converted to farmland and Earth would have lost three to four times as much forest.

Yet it's also true that synthetic fertilizers and other foundations of modern agriculture have warped global ecology, widened socioeconomic inequality, and degraded much of the world's soil. Many of the high-yielding crop varieties introduced in the Green Revolution depend on abundant water, fertilizer, and pesticides, disadvantaging farmers who cannot afford such resources. In some regions, the monoculture of high-yield staples has reduced the overall diversity of diets, especially among the poor. Leakage and runoff from excessive use of fertilizers and pesticides have polluted groundwater, lakes, and rivers; created algae blooms and dead zones in the ocean; and harmed pollinators and other wildlife. By some estimates, the Haber-Bosch process demands up to 2 percent of the world's annual energy supply and contributes 1.4 percent of global carbon dioxide (CO_2) emissions.

Haber seems to have recognized that his invention was neither wholly beneficial nor sustainable. "It may be that this solution is not the final one," he said in his 1920 Nobel lecture. "Nitrogen bacteria teach us that Nature, with her sophisticated forms of the chemistry of living matter, still understands and utilizes methods which we do not as yet know how to imitate. Let it suffice that in the meantime improved nitrogen fertilization of the soil brings new nutritive riches to mankind and that the chemical industry comes to the aid of the farmer who, in the good earth, changes stones into bread."

AFTER THE INITIAL EXCAVATIONS of our yard, Ryan and I reconsidered our plans for a garden. We were determined to improve the soil on our property in whatever ways were feasible and adapt to any constraints we could not overcome. As the heat and drought of summer eased, we realized that the overall situation was not quite as dire as we'd originally thought. When we dug a few small trenches in different areas of the backyard and filled them with water to test drainage, we discovered patches of relatively soft, dark, loose earth. Our garden soil was clearly not ideal, but it was also not universally terrible. My initial moment of despair, scrounging through the dirt on my hands and knees, may have been a touch melodramatic.

Our contractor, Ted, had already agreed to truck in enough high-quality soil to fill the raised beds; he offered to bring in even more to help amend soil throughout the yard. In early September, he and his colleagues arrived at our home with a slinger truck, essentially a dump truck equipped with a pivoting conveyor belt. They backed the truck into our driveway, extended the conveyor, angled it toward the backyard, and switched it on. Within seconds, a 30-foot-long stream of soil was arcing over the corner of our house and crashing onto our bare yard with all the force and drama of a waterfall. The soil, as dark and powdery as Dutch cocoa, billowed in the wind, coating our garage and fence in black dust and filling our nostrils with the scent of rich earth. In less than half an hour, the truck had deposited a modest hill of soil in our yard. Ryan and I spent several days wheelbarrowing most of it into the series of raised wooden beds that he had designed and built along the back fence. We spread the remaining soil throughout the yard, forming a new layer about three inches thick.

In order to cope with the underlying soil conditions, we resolved to deemphasize lush, resource-hungry plants in favor of hardy, drought-tolerant natives that thrived in, or at least tolerated, nutrient-poor soils, envisioning drifts of penstemon, hummingbird mint, and coreopsis bordering the flagstone path. As for the eastern half of the front yard, where I'd stripped away the turf and struggled to dig through the rubble-strewn dirt, I decided to try seeding a wildflower meadow. In contrast to perennials, many of the plants we call wildflowers have evolved to germinate, mature, and expire rapidly in impoverished and disturbed soils, producing copious seed to ensure a new generation.

These early adaptations to the challenges of gardening on our property were heartening, but they felt inadequate. Importing a thin layer of fertile earth was a superficial intervention—effectively a temporary balm. If we wanted to sustain a thriving garden on this landscape, we needed more than a rudimentary assessment of our soil's physical condition and a few basic amendments. We needed to understand soil on a much deeper level. Soil, I began to realize, was something I'd been surrounded by my whole life yet hardly knew. What *was* soil, exactly? Where did it come from? And how did one nurture it?

The soft, dark, fertile layers of earth that many people picture when they think of soil are a relatively recent development in the planet's evolution. The types of soil most familiar and important to us are dependent upon and inextricable from life. And for several billion years, there was no large or complex life on land. Our living planet typically requires centuries to create a single inch of fertile topsoil. Most of Earth's soils formed over tens of thousands of years, many over hundreds of thousands of years, and some over millions. Much like the fruits it would eventually nourish, soil needed time to ripen.

As soon as the planet's landmasses started forming around four billion years ago, wind, water, heat, and ice began, slowly but relentlessly, to disintegrate any exposed rock—a process known as weathering. Some of the resulting particles of rock remained more or less in place, whereas others were swept up by wind and running water and deposited elsewhere. Layers of crumbled rock weathered even further into primordial gray soils.*

The mineral particles in soil are usually categorized as gravel, sand, silt, or clay based on their size. Gravel is the largest, typically defined as loose rock between 2 and 63 millimeters in diameter. A grain of sand ranges from 0.05 to 2 millimeters, large enough to see with the naked eye and rub between one's fingers. Particles of silt, between 0.002 and 0.05 millimeters, are too small to observe without a microscope. And with diameters less than 0.002 millimeters, clay particles are by far the tiniest of all, about the same size as some bacteria.

When microbial life emerged, it undoubtedly began to alter the composition of Earth's first soils. Microbes chewed through rocks to extract their mineral components, turned those minerals into new compounds, and added carbon to the soil in the form of metabolic

* In 2018, geologists Nora Noffke and Gregory Retallack published the discovery of a 3.7-billion-year-old rock outcrop beneath a retreating ice cap in Greenland containing what may be the oldest fossilized soil on record.

byproducts and decomposing cells. Between 700 and 425 million years ago, more complex forms of life joined single-celled microbes on land: algae, fungi, lichens, and early terrestrial plants reminiscent of today's mosses, hornworts, and liverworts. The ancient four-limbed fish *Tiktaalik* has become the mascot of animals' journey from sea to land, but fossil evidence indicates that arthropods resembling millipedes and scorpions were the first animals to crawl out of the ocean and adopt a terrestrial lifestyle, perhaps as far back as 440 million years ago. Together, these pioneering land dwellers and literal trailblazers stimulated the formation of fertile, well-aerated soil by dissolving rock with acids and enzymes, digging burrows, and enriching mineral layers with their feces and residues.

By about 380 million years ago, forests covered much of the planet's land surface. The probing, acid-tipped roots of trees and shrubs fractured rocks, speeding up soil production while also protecting soils from erosion by anchoring them in place. Once microbes and fungi evolved the ability to digest tough plant tissues like cellulose and lignin, decomposing vegetation became one of the most important components of soil, reinvigorating it with essential nutrients. Grasses emerged as early as 100 million years ago and at one point covered 30 to 40 percent of the planet's land surface, creating the especially deep and fertile soils that eventually became some of the world's breadbaskets.

Earth's earliest soils were almost exclusively mineral, primarily composed of disintegrated rock and pores filled with air and water. In contrast, the majority of soils on the planet today are complex medleys of air, water, mineral particles, and organic matter, a broad technical term that encompasses living creatures and their carbon-rich secretions and remains. "The mixing of decomposed organic matter into the soil surface, which . . . became more pronounced after 400 million years ago, was a pivotal event in Earth's history," writes soil scientist Berman D. Hudson in *Our Good Earth: A Natural History of Soil*. "It was a necessary prelude to the establishment of the modern carbon cycle."

Teeming with multitudes of creatures from all of life's kingdoms—transformed by their continual activity and saturated with their residues—soil became both an immense reservoir of carbon, nitrogen, phosphorus, and other vital elements as well as a critical site of exchange, where those elements were free to move between the living and nonliving and cycle through rock, water, and air. One of the most important types of soil organic matter is humus: a handsome, dark, and mysterious substance whose precise composition is not yet fully understood but likely includes recalcitrant fragments of partially decayed cells, proteins, fats, and carbohydrates bound to mineral particles and bundled in soil aggregates. Whereas much of the organic matter in soil is consumed within days to decades, humus is more stable. Carbon-dating studies have demonstrated that some of the carbon in humus and other especially resilient forms of organic matter can remain in soils for millennia. Altogether, the planet's soils store somewhere between 2.5 and 3 trillion tons of carbon, which is around three times more than all the carbon in the atmosphere and about four times as much as in all living vegetation.

The continents are home to the great majority of living matter on the planet, much of which is concentrated in soil. Plants are the largest members of the soil ecosystem and its most important conduits, channeling water from land to sky as they pull atmospheric carbon into their bodies. Plants also feed a portion of the sugars and other organic compounds they photosynthesize to microbes and mycorrhizal fungi on and around their roots. In exchange, symbiotic microbes and fungi help plants absorb water and nutrients from the soil.

Alongside plants and relatively familiar creatures like ants, termites, and earthworms, all manner of weird and wonderful organisms populate soil, often clustering around root systems like savanna animals around watering holes. This underappreciated menagerie includes tiny, sometimes spectacularly colorful arthropods called springtails that can catapult themselves more than twenty times their own body length in a fraction of a second; oribatid mites, each about one-tenth the size of a lentil; shape-shifting blobs of amoebae called slime molds; transparent, ribbonlike nematodes, also known as roundworms; and microscopic animals called tardigrades, which resemble eight-legged gummy bears with hose nozzle mouths. Protozoans, a diverse group of single-celled creatures, move through films of water within soil's pores by contorting their gelatinous interiors and flapping their numerous appendages. A tablespoon of healthy soil easily holds a population of organisms many times the number of humans alive today. A single gram of fertile soil may contain billions of microbes and viruses, millions of protozoans and algae, hundreds of nematodes, dozens of mites and springtails, and a thousand meters of filamentous fungi.

Much like life, soil has thwarted all attempts to contain it within a succinct and precise definition. Most textbooks and scientific organizations rely on lengthy, convoluted definitions of soil that list its many properties and refer to it as a material or medium. There seems to be growing recognition in mainstream scientific circles, however, that soil may be best understood not as a substrate for life but rather as itself a living entity. *The Nature and Properties of Soils,* one of the most widely used soil science textbooks, states that after Earth's surface rocks came into contact with air, water, and life, they were "transformed into something new, into many different kinds of living soils," adding that soils are "living systems" whose diverse members "work together to function in a self-regulating and perpetuating manner."

Similarly, geologist Gregory Retallack, one of the world's foremost experts on the origin and evolution of soils, has written, "Both soil and life are complex interfaces, maintaining a dynamic equilibrium that is self-sustaining, taking and giving back materials to their environment." "The rhythms of annual leaf fall," he continues, "of decadal predator and prey fluctuations, and of millennial nutrient depletion and renewal are like the dance of muscle and nerve that create the rhythm of a heart-beat. . . . In some ways, life can be considered soil grown tall."

As I learned about the true nature of soil, I began to see our garden in a new way. I'd always thought of soil as a kind of earthly fabric—something to be conditioned—and as a repository of nutrients—something to be replenished. Now I began to perceive the soil in our yard as a living thing in and of itself. Simply amending our soil to make it loamier would not be enough.* If Ryan and I wanted to sustain

* Loamy soil, with an approximate ratio of 40 percent sand, 40 percent silt, and 20 percent clay, is considered ideal for gardening because it is easy to dig, well aerated, and free-draining.

a thriving garden on this landscape, we would have to tend to its long-term health. We needed to revive the soil ecosystem in our backyard.

With this new resolve, we set to work, focusing on interventions that would nourish our soil with organic matter, protect it from erosion, and increase the diversity of living creatures within it. To conserve water and reduce runoff, we installed a drip irrigation system. In the southwest corner of the yard, Ryan built a compost bin. Throughout the garden, we planted so many different species and cultivars of flowering perennials that we could not keep track of them without a detailed register. We added a heaping inaugural layer of mulch to the flower beds, a practice we intended to repeat at least once or twice every year. We began to leave the spent vegetation alone, allowing it to rot in place. Likewise, we raked most of the leaves that fell on our property onto the soil. After our last harvest, but before the first frost, we sowed the raised beds with a mix of winter-hardy legumes and grasses, such as vetch, crimson clover, peas, and oats, so that the soil would never be bare. In spring, we would mow them down and leave them to decompose.

Based on what I'd learned about soil science, the single most important change we made was introducing a large number of healthy plants to an emaciated landscape. If we'd somehow been able to peer beneath the surface of our garden and record a subterranean time-lapse, I think we would have witnessed a remarkable rebirth. As plant roots permeated the earth and loosened compacted soil, they became havens for microbes and fungi, awakening long-dormant processes of chemical transformation and nutrient cycling. Just below us, mycorrhizas were weaving new networks as earthworms, slugs, and arthropods processed large volumes of soil into durable, nutrient-loaded fecal pellets. Microbes, algae, and fungi were secreting sticky substances that bound tiny particles of soil into larger, more loosely spaced aggregates. By squirming, scuttling, and tunneling, worms, ants, and other animals were aerating soil, mingling its layers, and creating extensive networks of channels for roots to explore. Perhaps most fundamentally, microbes and fungi were decomposing the remains of all life forms, including other microbes, enriching soil with organic matter and replenishing its reservoir of essential nutrients. In spring, a tapestry of foliage would unfurl itself across the garden, further protecting the mulched soil from wind and weather. Carbon would once again flow rapidly from air to land through solar-powered lungs. Over time, the soil on our property would become softer, darker, and more fertile. Bit by bit, our soil was coming back to life.

AS A CHILD, ASMERET ASEFAW BERHE rarely gave much thought to soil. For most of her youth, her home country of Eritrea was fighting a war for independence. The front lines of the conflict constantly shifted, sometimes approaching the capital city of Asmara, where Berhe lived with her parents and several of her five siblings. She remembers bombs that shook every building and shattered windows across the region. There was not much opportunity to go hiking or play in the dirt. "As a young person, you didn't just go and wander around in nature," Berhe says. "There was a constant threat of land mines and other dangers." Although her house had a large garden, she seldom helped tend it, and she didn't regard soil as anything more than a medium in which to grow plants.

In 1991, when Eritrea was liberated, Berhe began attending the University of Asmara—one of only a thousand students in the country to begin college that year. At the time, she intended to major in chemistry, which had long been a favorite subject, with the long-term goal of becoming a medical doctor. While exploring her options, however, she became intrigued by an introductory course on soil science. For the first time it occurred to her that there was an entire dimension of the world that she had mostly overlooked: the very ground beneath her feet. In soil science classes, where she was one of just three women, Berhe learned about soil's eclectic and enigmatic composition and the staggering abundance of life within it. "Soil is the most complex biomaterial that we know of in the Earth system," she says. "There's simply nothing else like it."

She thought back to a drive her family regularly made from Asmara to the port town of Massawa. In Eritrea, the journey was famous for revealing "three seasons in two hours," as the saying went. The drive began in the relatively cool, semiarid climate of the capital city, situated on a plateau 7,600 feet above sea level. Winding mountain roads quickly descended into sub-humid regions with lush forests and terraced farmland, proceeded through much drier acacia woodlands, and finally arrived at a searing and nearly leafless desert along the coast of the Red Sea. The dramatic shift in scenery she had enjoyed on those family trips, Berhe realized, was not simply a product of varying altitude and temperature—it was also shaped by the reciprocal relationships between the living creatures and soils specific to each ecosystem. The environment determined what could grow, but over time, life changed its environment. Biology and geology, soil and climate—they were all bound up together.

After college, Berhe moved to the United States and eventually completed a PhD in biogeochemistry at the University of California, Berkeley, where she investigated how soil erosion alters carbon storage and exchange. As her career advanced, she earned one prestigious title after another. At the University of California, Merced, she became a professor of soil biogeochemistry and the Falasco Chair in Earth Sciences and Geology. She also served as chair of the U.S. National Committee for Soil Sciences at the National Academies. In May 2022, the Senate confirmed Berhe as the director of the Office of Science for the U.S. Department of Energy.

Over the years, the profound connection between soil and climate that Berhe began to comprehend in college became one of the central themes of her research. "Historically, the rate at which carbon was being taken up from the atmosphere through photosynthesis and stored in soil was roughly equal to the rate at which it was decomposed and released back into the atmosphere," Berhe explains. "But modern land use practices like deforestation, extensive tillage, and excessive application of chemicals have allowed less carbon to enter soil, while simultaneously increasing the rate at which that carbon gets released. The more we degrade the soil, the more we're skewing that balance."

In addition to eliminating native habitat and accelerating species loss, human modification of Earth's land surfaces—which encompasses deforestation, agriculture, and food production more generally, among other forms of land change—is responsible for around a third of global greenhouse gas emissions each year. These emissions include hundreds of millions of tons of methane from ruminants and rice paddies as well as millions of tons of nitrous oxide from manure and synthetic fertilizers, both of which warm the atmosphere much more than the same mass of CO_2. Over the past twelve thousand years, agriculture has resulted in the loss of 116 billion metric tons of carbon from the planet's soils, which constitutes about 17 percent of all the carbon humanity has released to the atmosphere.* In a self-amplifying feedback loop, climate change is exacerbating soil erosion and land degradation by swelling the seas and increasing the frequency and intensity of drought and extreme precipitation. "Climate and soil are intimate partners in a dance of millennia," writes biologist Jo Handelsman in her book *A World Without Soil*. "At its worst, the duo is destructive. . . . At its best, the duo is harmonious, improving soil health and stabilizing climate. Today humans are uniquely positioned to restore the duo to harmony."

Restoring the balance between Earth's skin, breath, and bones necessitates swift and sweeping transformations. One huge piece of such an undertaking is reviving and defending forests, grasslands, peatlands, wetlands, and other ecosystems, especially those with deep soils and thick vegetation. Another is changing the way the world produces, transports, and consumes food. Like existing energy infrastructure, modern agriculture and food systems require a revolution.

Numerous scientific, agronomic, and governmental organizations have proposed strategies to accomplish this reform, ranging from curbing meat consumption and food waste to genetically engineering supercrops. Many of the soil-focused strategies are modern applications of ancient agricultural methods. Over the decades, these practices have

* In climate science and policy, tons of carbon and tons of carbon dioxide (CO_2) are both commonly used metrics, sometimes leading to confusion. I have tried to be as consistent as possible within a given passage of the book. To convert between them, remember that one ton of carbon = 3.67 tons of CO_2. For example, 116 billion tons of carbon is equivalent to about 425 billion tons CO_2, which is 17 percent of the 2.5 trillion tons of CO_2 humanity has released to the atmosphere throughout history.

been bundled into various alternative approaches to conventional farming, such as conservation agriculture, climate-smart agriculture, and regenerative agriculture. Although some of these approaches are more clearly defined than others, they tend to have more in common than not, leaning on the same three core principles: minimizing soil disturbance, maximizing soil protection, and emphasizing diversity. The idea is to agitate soil as little as possible while sustaining a near-permanent cover of diverse living vegetation. Compared to traditional plow-heavy monoculture, these principles help maintain soil structure, increase organic matter, conserve water, enhance crop resilience, and support wildlife.

Reducing soil disturbance usually means leaving farmland untilled for long stretches or eliminating tilling altogether. Farmers who practice no-till or low-till agriculture often rely on selective herbicides to kill weeds without uprooting them. They also sow crops with seed drills, which cut thin slots in soil, sometimes through the remains of spent crops, and drop seeds into them. One of the most effective ways to shield agricultural soil from erosion is to plant cover crops, typically legumes and grasses that are sown in the fall, allowed to mature over winter and early spring, then chopped down and left to decompose, thereby replenishing stores of carbon, nitrogen, phosphorus, and other essential nutrients. Layers of compost and mulch likewise function as both physical armor and supplemental sources of organic matter. The practice of cultivating a wide variety of plant species and rotating through a diverse roster of crops season by season preserves soil nutrients, deters populations of pests and weeds from growing too large, and reduces the chances of losing an entire harvest to a single pathogen.

The Intergovernmental Panel on Climate Change (IPCC) estimates that a combination of ecosystem restoration and conservation agriculture, among other land-based interventions, could sequester between two and four billion tons of carbon each year. Similarly, renowned soil scientist Rattan Lal has calculated that improved stewardship of terrestrial ecosystems across the planet, including cropland and pasture, has the potential to capture and store 333 billion tons of carbon by

2100, returning the atmosphere to preindustrial levels of carbon dioxide. In many parts of the world, farmers are increasingly applying the core principles of conservation agriculture. As of 2017, 37 percent of U.S. cropland is managed with no-till farming, an increase of about 8 percent since 2012. During the same period, the use of cover crops in the United States increased 50 percent from 10.3 million to 15.4 million acres, although that is still only 5 percent of national cropland. Globally, farm acreage practicing conservation agriculture tripled between 2000 and 2019, from about 160 million to more than 500 million acres, comprising 14.7 percent of the world's cropland.

Although there is wide agreement in the scientific community that minimal tillage, cover crops, enhanced biodiversity, and other tenets of conservation agriculture have many benefits for soil, people, and wildlife, claims that these practices can sequester enough carbon to mitigate climate change are more controversial. Reliably quantifying carbon storage in soil over long periods of time is challenging, in part because the precise molecular compositions of humus and other forms of stable organic matter are difficult to discern.* Some experts contend that, due to various methodological problems, soil studies often fail to simulate conditions on the world's diverse working farms and ultimately underestimate the potential of soil to accumulate carbon. Recent findings suggest, however, that as the planet warms, some types of soil may lose even more of their capacity to store carbon than previously predicted.

There are also economic realities to confront. Conservation agriculture can be profitable in the long run by improving a farm's productivity and reducing the need to purchase large volumes of fertilizer, herbicides, and gasoline. But getting to that point requires a hefty initial investment of capital that many farmers do not have. Likewise, farmers who rent land are not always inclined to make such invest-

* Some studies suggest, for example, that reduced tillage does not always increase overall carbon but rather shifts carbon to higher layers of soil, where it is easier to measure. In certain situations, adding organic matter to soil can foment the activity of microbes that release CO_2 and nitrous oxide to the atmosphere. Yet other experiments have shown exactly the opposite effect.

ments, even if they can. The margins of error in farming are so slim that near-term profit often eclipses all else. In the succinct words of environmental writer Emma Marris, "Governments must pay farmers to build soil." As she explains, some countries "have already begun to move towards a model in which farmers are less independent business-people growing and selling food, and more government-supported land stewards managing a complex mix of food production, soil fertility, wildlife habitat and more."

In Africa, a crisis of land degradation has already forced dramatic shifts in agricultural systems. By the late 1970s, prolonged drought and impoverished soil had resulted in severe famine in the Sahel, a vast semiarid swath of grass, savanna, and woodland just below the Sahara stretching from Senegal in the west to Eritrea in the east. For generations, farmers had regularly hacked trees and shrubs on their property to mere stumps, which left their soil barren and vulnerable to the elements. In the early to mid-1980s, with guidance from Australian agriculturalist and missionary Tony Rinaudo, smallholders in Niger began regenerating and selectively harvesting the stunted forests on their land. In doing so, they rediscovered the many benefits of an ancient form of farming: agroforestry, the deliberate integration of trees and shrubs with crops or pasture. Trees stabilized soil, shielded crops from harsh winds, offered shade in times of extreme heat, and provided a convenient source of firewood and fodder for livestock. Leguminous trees such as acacia and their microbial partners also converted atmospheric nitrogen into more biologically useful forms. One species, *Faidherbia albida* or white acacia, was a particularly good crop companion due to its peculiar lifestyle: at the onset of the rainy growing season, it dropped its leaves and went dormant, fertilizing the soil and allowing light to reach crops. Studies have found that when millet is grown alongside white acacia, yields are almost twice as high. Over time, this particular type of agroforestry became known as farmer-managed natural regeneration.

Between 1975 and 2004, the number of trees in Niger's Zinder Valley increased more than fifty-fold. As of 2009, farmers in southern Niger were practicing agroforestry on 12.3 million acres and producing an additional five hundred thousand tons of cereal each year. Similar practices have now spread to Burkina Faso, Mali, Senegal, India, and Indonesia, among other countries. Sustainable land management specialist Chris Reij has described farmer-managed natural regeneration as "probably the largest positive environmental transformation in the Sahel, and, perhaps, in all of Africa." Similarly, a recent scientific review of regenerative agriculture concluded that, among all its associated practices, "agroforestry in its many shapes and forms perhaps has the greatest potential to contribute to climate change mitigation through [carbon] capture both above and belowground."

"Land degradation is a major crisis that a large part of the world has not woken up to," Berhe told me. "We need the soil system to continue providing food, fuel, and fiber and all the ecological services that keep our planet livable. But we can't expect that soil will continue to deliver if we keep extracting its resources without giving back. We know there are many ways to farm while preserving soil health and even sequestering a significant amount of carbon. Now we need to use them to reach a compromise."

AS I WRITE THIS SENTENCE, our garden in Portland is nearing its third year of life. Its metamorphosis has astonished us.

In late fall of 2020, I scattered a mix of wildflower seeds on the bare eastern half of the front yard and covered them with straw and chicken wire to protect them from birds and squirrels. Within weeks, the seeds were germinating, forming a dense mat of young plants. They survived the chill of winter and grew vigorously as the temperature rose. By April, the meadow was a sea of baby blue eyes. By mid-May, hundreds of blousy red poppies jostled in the wind alongside slender indigo cornflowers. In June, scores of pink Clarkia and ruffled yellow coreopsis brightened the palette.

Before we could plant the entirety of the backyard, we needed to give it a bit more structure. In addition to constructing the raised beds in the back, Ryan, who once worked for a landscaping company and had been teaching himself woodworking, designed and built a cedar fence with a central gate and rose arbor as well as trellises for table grapes, raspberries, and an espalier apple tree with multiple varieties grafted onto a single trunk. He also completed the massive jigsaw puzzle that was our winding flagstone path.

Meanwhile, I had been studying pond construction. We dropped a large plastic cattle trough into a hole Ted had dug for us and eventually filled it with two and a half feet of water and pots containing pickerelweed, arrowhead, water celery, forget-me-not, white star sedge, and a red-flowering water lily. On the advice of avid "ponders" on a web forum I had joined, I set about making a raised bog filter as well: a smaller, elevated tub cloaked in rocks and filled with pebbles and plants that thrive in marshy conditions, like rushes, cardinal flower, and society garlic. With the aid of a pump and buried pipes, water traveled underground from the pond to the bog and spilled back into the pond via a small waterfall. The constant circulation aerated the water as the tangle of plant roots in the bog filtered it, absorbing most of the nutrients and thereby limiting the growth of algae.

Across the path from the pond, we arranged rocks around a central boulder and interspersed them with plants adapted to hot and dry conditions, including succulents, rock roses, sea holly, fleabane, and lavender cotton. Near the front and center, we stationed an Arbequina olive tree, a variety that grows especially well in our climate, producing clusters of dark and aromatic fruit. In a long arc tracing the circumference of the rock garden, I scattered California poppy seeds, much as I'd done in the meadow out front. By its first summer, the rockery was a nonstop carnival, awash in color and forever thrumming with ecstatic bumblebees.

In the remaining open spaces, we primarily planted hardy, drought-tolerant species such as lavender, penstemon, hummingbird mint, Rudbeckia, coneflower, geum, scabious, and bellflower. In some of the cooler, less sun-seared areas, however, we included a few favorite plants that prefer moisture: columbines with golden-tailed flowers that streak through the air like shooting stars and intensely fragrant peonies that bloom in the most delicious shades of sunrise. Although we brought some more established plants from our previous home, more than half were new additions purchased from local nurseries in gallon pots or four-inch containers. Yet in just two years, most of them have grown to many times their original size. On opposite sides of the garden, more or less in line with the Arbequina olive tree, we planted giant mallows with hibiscus-like blooms and a crepe myrtle that fizzes with papery pink flowers in late summer. The mallows, no more than a foot tall when they went in the ground, are now seven feet tall and three times as wide as they started out; likewise, both trees have surpassed six feet.

Whereas I have focused on tending the meadow, pond, rockery, and flowering perennials, Ryan has assumed primary responsibility for the culinary section of the garden. From three central raised beds, he has coaxed tomato vine jungles and forests of corn; bushels of carrots, eggplants, and green beans; a cartload of pumpkins; a nearly year-round supply of kale and lettuce; and approximately enough zucchini to supply several national grocery chains. In pots spread throughout the garden, as well as in and among the bedded plants, we grow rosemary, thyme, parsley, and other herbs.

One of the greatest joys of the garden has been the wildlife it continually attracts. When we bought our home, the only wild creatures we saw in the yard with any regularity were spiders crawling through the grass. There was not much reason for animals to visit. By midsummer of its first year, the garden was whirring with insect activity. Some of the first newcomers I noticed were water striders dimpling the surface of the pond. Soon after, dragonflies and damselflies were warming themselves on the rushes and speeding through the garden in elastic arcs, sunlight glinting off their gossamer wings. Gray hairstreaks, cabbage whites, woodland skippers, and tiger swallowtails began to flit between the blooms. Birds now routinely bathe in the bog and forage in the underbrush. In fall, goldfinches descend on the seedheads of cornflowers, coneflowers, and black-eyed Susans. In winter, golden-crowned sparrows, dark-eyed juncos, and spotted towhees hop across the half-frozen pond to drink from the waterfall. Hummingbirds visit throughout the year, seeking sugary sustenance wherever they can find it, whether in the summer trumpets of an electric blue penstemon or the small mauve flowers dotting a spire of late-blooming rosemary. Mammals have appeared, too: Pacific jumping mice have made a home in the southwest corner of the garden, and a trail cam we installed has recorded a family of raccoons frolicking in the pond at night, squeezing their hands into every nook and crevice.

I have become particularly fond of the bees. Not just the honey bees, which are a domesticated European species, but also the many unique and oft-overlooked bee species native to North America: the sweat bees with iridescent emerald breastplates, the leafcutter bees that pad their underground nests with neat circles sliced from foliage, and the male wool carder bees that tirelessly patrol their chosen flower patches like sentinel drones.

Where once there was nary an ant, it is now difficult to dig even a shallow hole for a crocus bulb without uncovering a cocoon, a fungal filament, or several long, plump earthworms. But we can't really take credit for the revitalization of our land. Although we initiated the process, other forms of life performed most of the work.

We're under no delusion that our soil has been wholly transformed in under three years, however. There are still areas of gravel-strewn clay and borderline hardpan. Laboratory tests indicate that while our soil has surprisingly high levels of potassium and phosphorus—which are important for flowering—it still has a deficit of nitrogen. Maintaining a robust foundation of living roots and an ever-present mantle of vegetation, in combination with mulching, composting, and planting cover crops, will undoubtedly improve the soil's structure and fertility, but the process can proceed only so quickly, even with our assistance. Our gardening experience has not been without its challenges, conundrums, and failures, either. We have lost some of our most tender plants to heat, drought, ice storms, and unseasonable frosts. Vigorous grasses are starting to crowd the wildflower meadow. And raccoons have toppled and shredded the marginal pond plants so often that I have resolved to plant only deep-water species they can't harm as easily.

A garden is a perpetual negotiation. The garden we are helping create is not what the landscape would move toward without our involvement. If our land were left alone, it would soon be populated by a variety of native plants and so-called weeds and, given enough time, perhaps develop into something like the oak-studded prairies and savannas that covered this region centuries ago. Instead, we have filled it with an ark's worth of botanical diversity that would never have materialized here on its own. At the same time, our garden has diverged from our original vision in many ways as the idiosyncrasies of our property, and the life forms now inhabiting it, have pushed us to compromise and adapt. To garden is to engage in a form of coevolution—not just with plants and pollinators but also with roots and fungi, microbes and microfauna, sun and soil. Our garden is not ours alone: it is a collaborative and improvised performance by a motley ensemble of creatures, some whom we know well, others whom we never see.

The idea that Earth itself is a garden is one of the oldest metaphors on record, with numerous iterations in cultures around the world. The modern scientific understanding of our planet as a vast interconnected living system reframes the metaphor in an important way, however. Historically, especially in Western culture, the world has been portrayed as a passive garden: a preformed idyll over which we have complete dominion or a fertile but dangerous wilderness we must shape and tame. In myth and religion, the ultimate origin of Earth's bounty is often ascribed to a higher, external power or left unexplored. But for the great majority of its history, Earth was nothing like the relative paradise our species and so many others have enjoyed. And far from being passive, Earth and its constituent creatures are agents in their own evolution.

Earth is a garden that sowed itself, nurtured itself, and, through sentient life forms, eventually became aware of itself: a communal garden in whose creation and maintenance every member participates, consciously or not. Like many gardens, Earth has endured catastrophes over which it had no control; like many gardeners, Earth's creatures have inadvertently undermined the very system on which they depend, sometimes pushing it to the brink of collapse. On the whole, life and environment—garden and gardeners—have coevolved relationships that favor mutual persistence. These reciprocal bonds have made Earth amazingly resilient over spans of time so vast we cannot properly comprehend them.

For hundreds of thousands of years, our species has been learning how best to tend not only our personal gardens but also the planetary garden of which we are a part. Despite the gift and burden of our self-awareness, our progress has been anything but linear. Ancient wisdom has been lost, ignored, and rediscovered. Crisis has forced both error and innovation. Today, we know more than ever about the intricate ecological interdependencies that keep our planet alive. It has never been more important to apply that knowledge: to reject the idea that we are masters of the planet, while simultaneously accepting our outsized influence; to recognize that we and all living creatures are members of the same garden and embrace our role as one of its multitudes of stewards; to realize that our continued existence in this world is not a given. We are but one of innumerable organisms crawling along the skin of a living rock, wrapped in a film of air, whirling through the vacuum of space at unfathomable speed. The universe is indifferent to us, moving inexorably toward a state of maximum entropy in which living planets like ours—in which life of any kind—will be impossible. Earth is a beautiful rebellion and a precarious miracle: a garden in the void.

ONE MORNING, FOLLOWING a rare summer rain, I stepped into the garden. Small drops of water, smooth and luminous as crystal, clung to every leaf and twig. A mellow, earthy scent, like freshly dug beets, permeated the air. Formerly gray clumps of moss were now green and spongy. Near the pond, beneath the mallows, a robin flung leaves into the air, hunting for a meal.

As I knelt by the bog, checking milkweed for monarch butterfly eggs, I noticed the onionskin echo of a recently molted damselfly still stuck to the stem of a rush. Just below it, where the waterfall met the surface of the pond, bubbles formed and popped. Each was a tiny domed mirror in which I caught glimpses of my distorted reflection, the contours of trees and flowers, and the clouds in the sky. In each bubble, a different version of the garden; in each, one of many possible worlds.

WATER

CHAPTER

4

SEA CELLS

WHEN I ARRIVED AT WICKFORD HARBOR IN NORTH KINGSTOWN, Rhode Island, early one June morning, the sea was moderately calm with a distinct metallic sheen, like a wrinkled sheet of foil someone had tried to rub smooth. Vitul Agarwal, a young oceanographer, waved to me from beside a research trawler with the name *Cap'n Bert* painted on its hull. Dressed in jeans and a diamond-patterned sweater, Agarwal welcomed me aboard and introduced me to the captain, Steve Barber, whose gray hair spilled from the back of a baseball cap.

A few minutes later, we motored slowly into Narragansett Bay, picking up speed as we cleared the harbor. The sun was low and pendulous, dropping petals of light onto the water. Directly behind the boat, the sea churned shades of pear and crocodile. "I think we're going to find a lot out here today," Agarwal said, gesturing toward our frothing wake. "Because of the color?" I asked. He nodded.

It did not take long to reach our destination, one of the deepest parts of the relatively shallow region of the bay through which we were traveling, measuring only 21.5 feet. Every week since 1957, in one of the longest-running surveys of its kind anywhere in the world, scientists have come to this exact spot to study some of the most abundant and important life forms in the ocean: creatures so tiny that the vast majority are invisible to the naked eye, yet so essential to Earth's ecosystems that our planet would be virtually barren without them—creatures we call plankton.

Plankton, from the Greek *planktos* for "wandering" or "drifting," are a large and diverse collection of water-dwelling organisms that tend to flow with currents and tides. Nearly every liquid environment on the planet is home to plankton: the ocean, of course, but also rivers, lakes, wetlands, geysers, ponds, puddles, and even raindrops. Although they are defined by their tendency to drift, plankton are not completely passive; many move locally with impressive speed and vigor, and some make epic daily vertical migrations between the depths and shallows by adjusting their buoyancy. The total number of planktonic species, while unknown, is conservatively in the hundreds of thousands. Although most are less than an inch long and often microscopic, a few large animals also qualify as plankton because they are such listless swimmers. Bacteria and viruses populate the smallest end of the plankton spectrum. Certain jellyfish and their relatives, some of which are more than one hundred thirty feet long with their tentacles fully extended, inhabit the other. In between bobs a panoply of strange and wondrous creatures, many of which are little known and poorly studied—despite their power to change the planet.

Agarwal slipped on mint-green rubber gloves and picked up what looked like a comically large, incredibly fine-knit butterfly net missing its handle. A metal ring propped open the mouth of the net, while its narrow tail clutched a small plastic jar known as a cod end. "This is one of the samples we'll collect, concentrate, and preserve for the future," Agarwal said. "The goal is for water to go through the net and for things to get trapped in this little cod end. First what we want to do is get it to sink."

He lowered the net over the side of the boat with a rope and repeatedly dunked it into the water, the way one might dip a tea bag in a mug of hot water to weigh it down. The net billowed stubbornly near the surface. "Ideally, when there's a current—" Agarwal began to say, when the net suddenly straightened. "There we go. You see? It's going to stretch out." Soon the greater part of its tail had sunk out of view.

Agarwal prepared a few more nets, each of which had pores of a different size, ranging from 20 microns, about the diameter of a white blood cell, to 1,000 microns, roughly the size of a large grain of sand. Collectively, the nets would trap a diverse assemblage of minuscule organisms, some of which Agarwal would take back to the lab. After waiting for a quarter hour or so, he pulled one of the nets back onto the boat, removed the cod end, and poured its contents through a filter into another plastic receptacle. At first glance it looked like little more than water peppered with dust. As I peered closer, however, it became clear that the water was alive. The dark specks I mistook for dust were not merely floating—they were twitching. Other, tinier particles spun and sputtered. A few dime-sized jellyfish pulsated near the surface of the container, so diaphanous they seemed to phase in and out of existence with the shifting light.

"Now I'm going to concentrate this entire thing into that," Agarwal said, pointing to a glass container that resembled a small jam jar. He carefully poured the sample from one vessel to another, straining it through a series of filters. As he worked, he put aside most of the clear water that passed easily through the filters and focused on the murkier fluid that was left behind. The process reminded me, once again, of brewing tea—in this case, loose leaf—except that the goal was to savor the dregs and discard everything else.

By the time Agarwal finished concentrating the sample in the small glass jar, it had developed the honeyed hue of apple cider. Thousands of tiny creatures—shaped like discs, rowboats, and boomerangs—were moving of their own volition. Some leapt through the water, flea-like, almost teleporting from one position to another. Others glided along as liquidly as manta rays or bored ahead as though excavating a tunnel. Many of those energetic motes were likely miniature crustaceans known as copepods, Agarwal told me, and they constituted a fraction of the life in the vial. The fluid, amber and cloudy, was full of living things too small to discern without a microscope. "For every plankton you can see, there are at least ten—maybe a hundred—that you can't," Agarwal said. "And this is just one sample." He looked at me with wide eyes, then out at the sea. "Now think about how much life there is in the water."

IN OUR PLANET'S FIRST half billion years or so, as torrential rains submerged incipient landmasses, Earth was a true water world, entirely covered by the early ocean, save for a few volcanic islands. The ocean still covers more than 70 percent of the planet's surface today and contains more than 96 percent of all the water on Earth. At first, the ocean may not have been particularly salty. Over time, rain, wind, ice, and surf weathered the thickening continental crust, freeing minerals and salts such as sodium and chloride ions, which flowed to the sea. When sea water evaporated, the salt remained behind, steadily accumulating.* The ocean, then, is a hybrid brew, partly atmospheric and partly terrestrial in its composition. The ocean is Earth's great cauldron, its mixing vessel, where the planet's three main spheres converge and their elements coalesce.

This possible scarcity of salt is only one of the many ways the ancient ocean would have seemed peculiar to us. Just as the continents and atmosphere are to some extent biological constructs, many of the ocean's defining features throughout Earth history were the result of the life it harbored. Although single-celled marine organisms evolved relatively soon after the planet's formation, it took several billion years for much larger and more complex creatures to emerge. In the interim, different groups of multiplying microbes may have, at various times, dyed parts of the sea green, rust red, pinkish-purple, black, or milky white as they and their metabolic byproducts reacted with the ocean's primordial chemistry. About 530 million years ago, during the Cambrian explosion, the first fish began to populate the sea, revolutionizing marine food webs. But it would take even longer for life to help establish what we recognize as modern ocean chemistry—a fundamental transition on which all Earth life would come to depend. The most important living participants in this global transformation were not

* Average ocean salinity has changed through Earth history and varies by geography, yet for reasons that are still being worked out in full, the many different processes that add and remove ions to the ocean are more or less in balance today, maintaining the ocean at its current level of saltiness.

fish or other relatively large and iconic ocean creatures but rather the smallest and humblest among them: the plankton.

Before making the trip to Rhode Island, in an effort to become better acquainted with the ocean's tiniest citizens, I spent many happy hours gazing at photos of plankton, ensorcelled by their beauty. Like larger and more familiar sea creatures, plankton often rely on shells or skeletons for support and protection. The sheer diversity and sculptural intricacy of these structures is staggering, far surpassing any sand dollar, scallop, or conch. Viewed up close, some plankton look like chandeliers, wicker baskets, or spun sugar confections. Others resemble the webbed sails of windmills, wheels of citrus, or bits of ribbon candy. Still others call to mind pinecones, harpoons, knitting needles, meandering golf tees, inverted mushroom caps, slivers of rainbow, and fireworks frozen mid-burst. Inspired by this kaleidoscopic beauty, a few nineteenth-century naturalists created exquisite mosaics and mandalas by arranging jewel-like plankton on glass microscope slides, painstakingly positioning each one with a single hair from a horse or boar. These miniature marvels commanded high prices from collectors and delighted guests in Victorian salons. Although the modern, high-resolution photographs I perused were breathtaking, I, too, wanted to experience plankton's phantasmagoria firsthand, which meant I would need access to a fairly powerful microscope—and someone who knew how to use it.

The afternoon after sampling plankton in Narragansett Bay, I met Agarwal at the University of Rhode Island's Graduate School of Oceanography, which was just steps from the shore. I found him huddled over a microscope in the lab where he spent many long hours every week counting and identifying plankton. Beside him were several well-worn field guides with detailed sketches of local plankton and a tally counter with long rows of oversized plastic buttons, each labeled with a different species, which helped him understand how the composition of the bay's plankton population changed over time. Behind us stood part of the university's six-decade plankton archive. Agarwal opened one of the boxes and pulled out some glass vials, whose fluid contents ranged in color from saffron to walnut to moss green. Each vial held a sample of plankton suspended in iodine, which preserved their cellular structures so that they could be examined long into the future.

Agarwal invited me to sit in front of the microscope and observe some of the plankton we had collected earlier that day. A nearby monitor connected to the microscope displayed the magnified images, allowing us both to see the same image simultaneously. As I adjusted the knobs for focus, a long segmented creature bristling with spines came into view, reminding me immediately of a house centipede. "That's *Chaetoceros,*" Agarwal said. Each segment, he explained, was a single-celled photosynthetic plankton with a spiky silica shell that had joined this chainlike colony. Other plankton in the slides we inspected looked like thinly sliced almonds, dumbbells, overwrought Christmas-tree stars, and martini olives skewered on a toothpick. Eventually we encountered a plankton that resembled an extremely delicate icicle. "This long, needlelike thing is a phytoplankton, probably another diatom," Agarwal said as he flipped through a yellowed pamphlet titled *Guide to the Phytoplankton of Narragansett Bay, Rhode Island,* trying to match the species. "Life is . . ."—he paused—"complicated."

Because the term *plankton* is a catchall for an extremely diverse assortment of drifting organisms spread across the tree of life, rather than a discrete cluster of closely related species, scientists have sorted and slotted them into numerous overlapping classification systems. Broadly speaking, plankton fall into two big categories—the plantlike phytoplankton and the animallike zooplankton—though quite a few species have characteristics of both. Cyanobacteria and other microbial, ocean-dwelling phytoplankton are Earth's original photosynthesizers. About half of all photosynthesis on the planet today occurs within their cells. Despite their ubiquity, phytoplankton still preserve many mysteries. It was not until the 1980s that oceanographers Sallie "Penny" Chisholm, Robert Olson, and their colleagues brought a laser-equipped cell counter out to sea and discovered a species of cyanobacteria called *Prochlorococcus,* which turned out to be the tiniest and most abundant photosynthetic creature on the planet. There are an estimated twenty thousand *Prochlorococcus* cells in a drop of seawater—

and three octillion on Earth—yet they were so minute that no one had noticed them before.

Single-celled algae known as diatoms comprise another widespread group of phytoplankton. Diatoms have glass exoskeletons: they encase themselves in rigid, perforated, and often iridescent capsules of silica, the main component of glass, which fit together as neatly as the two halves of a cookie tin. A different group of microalgae, the coccolithophores, also sheathe themselves in armor—made not of glass but of chalk. They construct shells out of overlapping scales of calcium carbonate, the mineral from which limestone and marble are composed, and which was once commonly used to write on blackboards.*

Just as plants form the base of the food chain on land, phytoplankton nourish the seas. Zooplankton eat their green cousins as well as each other. Radiolarians are single-celled zooplankton that, like diatoms, produce glass skeletons from silica. Their armor is typically conical or spherical, trellised, and adorned with curious spikes and projections, evoking baroque thimbles and ethereal Sputniks. Foraminifera use sand, silt, calcium carbonate, and even the remains of other plankton to construct chambered shells in a variety of shapes: open-ended tubes, nautilus-like whorls, bunches of what look like lychees. Unlike most single-celled plankton, foraminifera can grow surprisingly large, sometimes more than seven inches long. Tintinnids, a name derived from the Latin word for "jingling," live in bell-shaped shells, using a wreath of mouth bristles to catch smaller microbes. Dinoflagellates, which often resemble spinning tops, twirl through the water using ribbon and whiplike appendages and shield themselves with plates of cellulose, the same organic compound that gives the walls of plant cells their rigidity. About half of all dinoflagellate species eat other microorganisms, while the other half are photosynthetic. When jostled, some glow an otherworldly blue, illuminating tumbling waves, the flanks of whales and submarines, and fresh footprints in the sand.

The smallest plankton are consumed by larger plankton, including

* Modern classroom chalk and sidewalk chalk are usually made of gypsum.

the larvae of fish and crustaceans, which in turn feed a succession of bigger sea creatures, from herring and squid to seals and dolphins, so that plankton ultimately support all marine life. Some baleen whales, the largest animals that have ever lived, survive exclusively on a diet of tiny fish, krill, and plankton—a testament to the abundance and significance of the minuscule. A single drop of seawater might contain tens of thousands of plankton on average, but at times it will hold many more. When storms or shifting winds and currents transfer a surplus of deep, nutrient-rich water to the surface or rivers dump agricultural and residential fertilizers along the coast, certain types of plankton—namely dinoflagellates and diatoms—multiply much more quickly than usual, potentially crowding every fifth of a teaspoon of water with millions of cells. These plankton blooms, which are sometimes visible from the stratosphere, can swell to an astonishing 770,000 square miles, an area roughly the size of Mexico. Plankton are so tiny and ubiquitous, they sometimes seem less like creatures within the ocean than atoms of the ocean itself. Without plankton, the modern ocean ecosystem—the very idea of the ocean as we understand it—would collapse.

JUST ACROSS THE HALL from the lab where Agarwal counts plankton is the office of his graduate advisor, Susanne Menden-Deuer, a plankton ecologist and professor of oceanography. On the first day of my visit, Menden-Deuer met me in the parking lot, wearing maroon corduroys and a gray cardigan, her blond hair pulled back into a neat braid. She showed me to her office, which was decorated with an illustration of copepods by Ernst Haeckel, maps of Puget Sound—where she used to live and work—and a couple of succulents with Rapunzel-like strands of leaves cascading down her desk.

Even as a child, Menden-Deuer demonstrated an intense curiosity about the living world: "My older sisters tell me I wouldn't eat a pea without opening it because I had to see what was on the inside," she recalled. While studying biology at the University of Bonn in Germany in the 1990s, she received a stipend to visit the University of

New South Wales in Sydney, Australia, and pursue marine science, which had always fascinated her. In Sydney, she lived a short stroll from the beach, where she frequently snorkeled, observing the occasional congregation of hypnotically colorful sea slugs known as nudibranchs. She got a job as a research diver, feeding wild sea urchins as part of an experiment, and joined the student dive club, through which she met her future wife, Tatiana Rynearson. "We fell in love diving," Menden-Deuer told me. In a single buoyant, sun-kissed year, her whole life seemingly merged with the sea.

After earning their PhDs in oceanography, the couple struggled to find faculty positions at the same university or even a job that would allow them to share benefits. Eventually, they both secured professorships at the University of Rhode Island, where they have worked ever since. Their offices are just a few doors apart.

Over the course of her career, Menden-Deuer became increasingly interested in the hidden connections between the minute and the colossal—between the fluttering of a single cell and the rhythms of an entire planet. "My approach is to recognize that individual plankton species matter and small-scale interactions matter," Menden-Deuer said in one of our conversations. "But what is important at the end of the day is the global scale. What drives my research is: How do we measure these small-scale processes, and how do we link them to the big picture? Planet Earth is one system and everything is interconnected in that system. Plankton are key players in how elements move through Earth and keep moving along. They are literally the engines that make biogeochemical cycles work. Plankton make Earth habitable. They have been making the planet habitable for billions of years."

We often hear how plankton do the opposite: apart from being whale food, one of the most familiar facts about plankton is that in large numbers they can poison the sea. Plankton blooms, also known as algal blooms or red tides—though they can also turn the sea orange, yellow, brown, or pink—sometimes suffuse water and air with toxins that sicken and kill fish, shellfish, birds, and mammals, including humans. As the prolific microalgae begin to die, the microbes that decompose their cells consume much of the available oxygen in the region, suffocating other creatures and creating what's known as a dead zone. In many cases, however, a modest plankton bloom neither produces toxins nor grows rapidly enough to deprive other organisms of oxygen; instead, it becomes a welcome smorgasbord. Dramatic as it may be, plankton's dual role as blight and bounty suggests but a sliver of their full significance. Through their growth and behavior, their life and death—their very presence—plankton modulate the chemistry of the seas and ultimately the planet as a whole.

In the 1930s, American oceanographer Alfred Redfield observed that the average ratio of nitrogen and phosphorus in samples of water collected from the deep ocean around the world was the same as the average ratio of these elements in the cells of phytoplankton: sixteen to one. Based on decades of research, Redfield eventually argued that plankton "not only reflected the chemical composition of the deep ocean, but created it," as biological oceanographer Paul Falkowski has phrased it. As dead plankton sank into the deep sea, Redfield proposed, bacteria decomposed them into their chemical constituents, enriching the ocean depths with the exact same proportions of nitrogen and phosphorus. Plankton, he elaborated, also maintained the ratio of these elements by continuously converting nitrogen into different chemical forms as part of ecological feedback loops, similar to those that microbes orchestrate on land.

When the marine reservoir of nitrogen dwindles relative to phosphorus and many nutrient-starved organisms begin to struggle, specialized nitrogen-fixing microbes thrive, adding ammonia and other biologically useful forms of nitrogen to the sea and ultimately replenishing the reservoir. Should nitrogen levels rise too high, other types of plankton, enjoying the glut, outcompete the nitrogen-fixing microbes. At the same time, increasing numbers of plankton transport more carbon to the oxygen-deprived depths as they die and sink, stimulating the growth of microbes that convert ammonia back to gaseous nitrogen as part of their respiration, which further stabilizes the ratio of nitrogen and phosphorus.

Since Redfield's day, scientists have discovered that the mechanisms behind this oceanic homeostasis are far more complex than he initially envisioned and that the proportions of elements in the ocean vary in more nuanced ways than he could have known, especially at the local level. Nonetheless, numerous studies have confirmed Redfield's primary insights and the existence of what is now called the Redfield ratio, though the precise processes responsible for this chemical balance are arguably some of the most important mysteries in oceanography.

Plankton are also a crucial component of both short- and long-term processes that sequester carbon and regulate global climate. Throughout its history, Earth has endured repeated periods of widespread glaciation that extinguished many species and severely inhibited life in general. Yet each time, our planet not only recovered but, eventually, flourished. How? The onset and thaw of ice ages are partly controlled by the shifting positions of continents and variation in ocean currents, which redistribute heat across the globe, as well as by changes in Earth's orbit, wobble, and tilt, which alter the amount of sunlight it receives. In some cases, however, our living planet's own self-stabilizing processes come into play. This resilience hinges in part on the exceptional versatility of that abundant and gregarious element from which all Earth life is made: carbon. Carbon's circular journey through air, land, and sea—its perpetual shuttling between organism and environment—ultimately acts as a planetary thermostat.

Carbon dioxide in the atmosphere continuously dissolves into the ocean's surface, where sun-loving phytoplankton incorporate it into their cells during photosynthesis. Much of this carbon is released in shallow waters when zooplankton and microbes eat and decompose phytoplankton, consuming oxygen and exhaling carbon dioxide in the process. Phytoplankton that evade consumption usually live for days or at most weeks. When they die, they bump into each other, form little clumps, and begin to sink, along with the fecal pellets of zooplankton, carrying carbon to deep, cold, dense water, where it may remain for thousands of years.* Some of this perpetual underwater precipitation, known as marine snow, feeds deep-dwelling creatures, but a portion continues to sink and settle on the seafloor, accumulating

* Yes, even plankton poop.

in layers of muck that eventually petrify and trap carbon for millions of years.

In parallel, carbon dioxide spewed by volcanoes combines with water vapor in the atmosphere, forming carbonic acid that falls to land in rain. Due to its slight natural acidity, rainwater reacts with and dissolves the planet's crust. The chemical reactions involved in this weathering produce various minerals, salts, and other molecules, which flow to the ocean via rivers, nourishing marine life. Certain types of cyanobacteria, plankton, corals, and mollusks use calcium and bicarbonate ions produced by weathering to construct shells, sheaths, skeletons, reefs, and stacked microbial mats called stromatolites. When such creatures die, their carbon-rich remains gradually accumulate in layers of compacted limestone sediment on the seafloor. Over great spans of time, tectonic activity subsumes and transforms the sediments, returning the carbon they contain to the planet's surface in the form of new mountains or erupting volcanoes, thereby completing the cycle.

If Earth enters a torrential hothouse state, intense and frequent rainfall weathers rock more quickly than usual, flooding the ocean with minerals, nourishing life in the sea, and removing carbon from the atmosphere faster than volcanoes can replenish it. Over hundreds of thousands to millions of years, this feedback loop cools Earth. Conversely, if ice smothers most of the sea and land, the water cycle effectively stalls, the productivity of plankton and other ocean life drops, and carbon dioxide builds up in the atmosphere, eventually warming the planet. "This entire process is therefore both largely controlled by life and ultimately allows life to exist on Earth," write paleontologist Peter Ward and geobiologist Joe Kirschvink. Although some of Earth's self-stabilizing processes can operate abiotically, life has been thoroughly entangled with the carbon cycle and planetary thermostat since its emergence more than 3.5 billion years ago.

Scientists have estimated that if phytoplankton vanished, the amount of carbon dioxide in the atmosphere would double, reaching levels the planet has not experienced since the early Eocene Epoch 50 million years ago, when the average global temperature was about 14.4°F higher than today and crocodiles swam in the Arctic. Conversely, if all the plankton-populated, high-nutrient regions of the ocean were maximally productive, atmospheric CO_2 would halve, plunging past preindustrial levels into a new ice age.

Planktonic confetti and other forms of marine snow accumulate on about 60 percent of the seafloor today. The uppermost layers of these sediments are like slurries, almost fluffy in texture, explains micropaleontologist Paul Bown of University College London; a few feet down, as the pressure increases, squeezing out water, they develop the consistency of toothpaste. Eventually, they are compressed into rock and are either melted in Earth's interior or returned to the surface by, say, clashing continental plates or shrinking seas.

If you chip off a piece of the White Cliffs of Dover and examine it with an extremely powerful microscope, you will see a jumble of granular detritus. Look carefully and distinct shapes will start to emerge: bows and discs made of tiny bonelike pegs packed together as neatly as the wedges in a stone archway. If you're extremely lucky, you might even find a relatively intact sphere of ribbed discs still clinging to one another, like a bundle of petrified doilies. You would see these things because the White Cliffs of Dover are more than just rock—they are also fossils. The cliffs' mineral building blocks—those intricate arcs, discs, and spheres visible only with a microscope—are the husks of single-celled coccolithophores that lived during the Cretaceous Period between 145 and 66 million years ago.

In fact, the vast majority of chalk and limestone formations on Earth, including large sections of the Alps, are the remains of plankton, corals, shellfish, and other calcareous sea creatures. Every imposing edifice that humans have constructed with limestone, including the Great Pyramid of Giza, the Colosseum, Notre Dame, and the Empire State Building, is a secret monument to ancient ocean life. And coccolithophores are not the only plankton that turn to stone, either. Millions of years ago, early tool-making humans discovered the benefits of working with flint and chert, which, unlike most rocks, are simultaneously hard, sharp, and knappable. Though they had no way of knowing, they were crafting arrows and axes from the compacted husks—the glass ghosts—of diatoms and radiolarians. Stone tools revolutionized our ancestors' diets, cultures, and technologies, which is to say that the mere remnants of plankton defined the course of human evolution.

Like plankton, many other sea creatures build skeletons or shells from calcium carbonate, including shellfish, sea snails, nautiluses, urchins, and corals, which are colonial organisms composed of symbiotic microbes, algae, and tiny gelatinous animals called polyps. Of all these calcifying organisms, free-swimming plankton have by far the greatest influence on the planet. Before the evolution of coccolithophores and other limestone-armored plankton, the movement of carbon and calcium through the ocean was very different. Limestone deposits were much smaller and were restricted to shallow continental shelves where corals thrived. Between 200 and 150 million years ago, however, calcifying plankton evolved, filled the open ocean, and became a critical link in the long-term carbon cycle that ultimately governs global climate. In parallel, they created new massive beds of limestone on the deep ocean floor, which stabilized the ocean's chemistry during times of crisis. Throughout Earth's history, intense volcanic activity has occasionally belched huge amounts of carbon dioxide into the atmosphere, which then dissolved into the ocean, dramatically acidified the water, and contributed to some of the worst mass extinctions on record. The limestone beds formed from dead plankton counteracted this process to an extent by dissolving in acidifying waters, releasing carbonate ions and raising pH, thereby protecting ocean life.

The unprecedented rate at which our species has flooded the atmosphere with carbon over the past few centuries may severely impair this natural buffer, however. The oceans are already 30 percent more acidic on average than they were in 1850; by the end of the century, their acidity may double, with devastating consequences for global ecology.

Ocean acidification disrupts a wide array of biological processes in numerous species, including metabolism, reproduction, embryonic development, and predator detection. Acidification even seems to alter the acoustic properties of seawater, interfering with echolocating dolphins and whales. The creatures that suffer most directly are the ones that form the foundation of the ocean's food webs. The chemical reactions between carbon dioxide and seawater diminish levels of calcium carbonate, making it much more difficult for plankton and other calcifying creatures to construct their shells and skeletons. When the pH drops too low, these organisms literally begin to dissolve. Plankton nourish all other marine life; corals alone support 25 percent of marine biodiversity. If global warming and ocean acidification proceed at their current pace, populations of calcifying plankton around the world will deteriorate and disappear; tropical coral reefs as we know them will likely collapse before the end of the century, replaced in some regions by mucilaginous carpets of algae and sponges; international stocks of salmon, tuna, mackerel, cod, herring, crab, lobster, shrimp, oyster, mussel, scallop, and clam will decline; and the global ocean ecosystem may shrivel to a thin and sickly version of its former self—a comparatively desolate expanse unlike anything in the past 60 million years. Even if all carbon emissions ceased by 2100, it would likely take somewhere between tens to hundreds of thousands of years for ocean chemistry to restabilize and for life to recover.

Plankton change the sea above our heads, too. Some plankton produce a sulfurous compound in their cells called dimethylsulfoniopropionate (DMSP), which may protect them from freezing temperatures, ultraviolet radiation, and fluctuations in salinity. When plankton die, DMSP seeps into the sea, where microbes break it down, producing the gas dimethyl sulfide (DMS). As DMS rises into the atmosphere, it reacts with oxygen to form sulfate aerosols that seed rain clouds. This interaction is of particular importance over remote stretches of ocean far from land, where soot, dust, and other terrestrial cloudseeding particles are scarce.

In 1987, James Lovelock and several other scientists published a paper arguing that the connection between ocean plankton and clouds might temper global climate. Named for its originators (Robert **C**harlson, James **L**ovelock, Meinrat **A**ndreae, and Stephen **W**arren), the CLAW hypothesis proposes the following feedback loop: when the temperature of the sea surface or the amount of sunlight reaching it increases, plankton thrive and produce more DMS, which stimulates cloud formation, which has the effect of reflecting more sunlight, cooling the planet, and slowing the growth of plankton. Years later, Lovelock expanded the concept to include another possibility: if the temperature of the ocean gets too high, the physical processes that bring deep, nutrient-rich water to the surface might fail, restricting the growth of plankton, reducing cloud cover, and ultimately exacerbating global warming. Although these highly simplified models remain controversial and undoubtedly overlook many nuances of the ecological system they describe, relatively recent studies have confirmed that they are at least partially correct. A comprehensive explanation of the relationship between plankton, clouds, and temperature—and the degree to which it matters for global climate—is still under active investigation.

Plankton are also the secret composers of the coast's most beguiling features, such as sand, seafoam, and the smell of sea air. Some of the loveliest beaches in the world, including the pink sands of Horseshoe Bay, Bermuda, owe much of their complexion to the colorful shells and skeletons of plankton. When plankton blooms dwindle and perish, the wind and waves often mix their decomposing proteins and fats with other bits of organic detritus, such as fragments of coral, seaweed, and fish scales. This moldering mélange acts as a foaming agent, generating numerous air bubbles that balloon into a thick froth, a kind of plankton meringue, which washes onto shore. Meanwhile, the sulfur aerosols generated by dying and decomposing plankton—the same ones that seed clouds—give sea air much of its characteristic funk, an odor reminiscent of boiled beets. That scent mingles with briny bromophenols, produced in large quantities by marine worms and algae, and the strong "ocean smell" of certain seaweed sex pheromones. On a sterile planet, the seaside would not smell like the sea—at least not as we know it. It might not smell like much of anything. When you breathe in sea air, you are literally breathing in sea life.

Because plankton are so ubiquitous, tiny, and easily dispersed, their orbit of influence extends far beyond the ocean and coasts. Every year, the wind carries immense quantities of Saharan dust across the Atlantic Ocean, depositing 27.7 million tons—enough to fill more than a hundred thousand semi-trailer trucks—in the Amazon rainforest, where it provides trillions of plants with iron, phosphorus, and other essential nutrients. This fertilizing dust is not simply tiny bits of dirt and rock; it is largely composed of the skeletons of ancient diatoms. Much of it comes from the Bodélé Depression, a sunbaked bowl of sand that was once the bottom of an enormous lake larger than all of North America's Great Lakes combined. Long after they have died, plankton continue to shape and sustain the planet, circulating vital elements through ocean, desert, and jungle. In their eon-spanning metamorphosis—their transformation from floating cell to entombed rock to windswept dust and back again—they embody the reciprocity of life and environment and the perpetual reincarnation of Earth.

TOWARD THE END OF my visit to Rhode Island, I asked Menden-Deuer to show me some of the most unique and beautiful plankton she had sampled over the years. She opened a file on her computer with hundreds of stunning photos, some of which had graced the covers of research journals. We marveled at images of diatoms and dinoflagellates that looked like tuning forks, dandelion seedheads, strings of jade beads, and Alexander Calder's kinetic sculptures. Menden-Deuer was particularly fond of a plankton that resembled the segmented circumference of an ornate clockface.

"This is a fun species," she said. "It's called *Eucampia zodiacus*—*zodiacus* for the zodiac. It's actually a 3D spiral, but when people first looked at it under a microscope, they just saw a flat shape, so they attributed that circularity."

"So it's really a Slinky?" I said.

"Yeah, exactly."

Menden-Deuer showed me an art installation on which she had collaborated with new media artist Cynthia Beth Rubin. In 2016, as part of the Open Sky Gallery, films created by dozens of artists around the world were displayed on the 828,000-square-foot LED screen of Hong Kong's tallest skyscraper, the 108-story International Commerce Centre. The video that Beth Rubin and Menden-Deuer submitted, which received an honorable mention from the judges, was a dreamy black-and-white pastiche of krill, jellyfish, and microscopic plankton, most of which were documented during research expeditions to Antarctica. For several minutes on a May evening, the silhouettes of these tiny ocean creatures drifted, skittered, and pulsed across the surface of one of the tallest structures humans have ever built, drawing eyes in a city of more than seven million.

The exhibit reminded me of a surprising fact I had recently learned about the 1900 Paris Exposition (Exposition Universelle), the seven-month-long world's fair intended to showcase and celebrate the ingenuity of modern civilization. More than 50 million people visited the exposition, where they rode a Ferris wheel, a moving sidewalk, and an escalator; watched motion pictures with sound; and admired the hulking steam-powered generators behind the incandescent Palace of Electricity.

Paris commissioned a relatively unknown architect named René Binet to design the Monumental Gate (Porte Monumentale), which would designate the entrance to the exposition and house the ticket booths on one of the city's main public squares. Binet's gate consisted of a gigantic honeycombed dome perched atop several grand archways, the foremost of which culminated in a lobed spire on which stood the statue of a woman, dressed in contemporary Paris fashion, her arm outstretched to welcome fairgoers. Constructed primarily from iron and plaster, the gate was covered with decorative stone, Byzantine motifs, and colorful glass cabochons. Thousands of blue and yellow lights illuminated the structure from within.

The architecture exuded grandeur and opulence, evoking a formal display of crown jewels, yet it was also delicate, airy, and distinctly organic. One writer of the era saw "the vertebrae of the dinosaur in the porch, the cells of the beehive in the dome and corals in the pinnacles." But none of these creatures was the primary inspiration for Binet. His true muse was much more obscure. As he designed the Porte Monumentale, Binet routinely visited libraries in Paris to study illustrations by German scientist Ernst Haeckel. Today, Haeckel is best known for his vivid and captivating drawings of animals, plants, and fungi, especially those collected in his wildly successful book *Kunstformen der*

Natur ("Art Forms in Nature"). His meticulously observed illustrations, arranged with an artist's eye for pleasing symmetry, have been reproduced countless times in every imaginable form, from murals, wallpaper, and framed prints to T-shirts, tote bags, and shower curtains.*

Haeckel was enamored of sea creatures, such as sponges, jellyfish, and their relatives, which were the focus of his first several monographs. He especially liked the elaborate yet precise geometry of radiolarians, which appealed to his exacting aesthetic. He would often collect the plankton himself and sketch their anatomy for hours, his left eye trained on a microscope as his right eye guided his hand. These were the images that obsessed Binet. "At present, I am building the Monumental Entrance for the Exposition of 1900," he wrote to Haeckel in 1899, "and everything, from the general composition up to the smallest details has been inspired by your studies."

If the limestone facades of the Great Pyramids and Notre Dame were secret monuments to plankton, here was an explicit one. Binet's organic sculpture of stone, metal, and glass was a tribute to evolution—in particular, its power to produce astoundingly complex and beautiful structures that rivaled, and often transcended, human design. Given what we now know about the importance of plankton to global ecology, these towering arches—a literal gateway to a celebration of human achievement—take on new meaning. A plankton expanded into a cathedral allows what is normally unseen to mesmerize, what is usually silent to reverberate. *Without me,* it seems to say, *you would not be here. Without me, none of this would be possible.*

If plankton had not infused the sea and air with oxygen, modulated ocean chemistry, and become key regulators of global climate, there would never have been forests, grasslands, or wildflowers, nor dinosaurs, mammoths, and whales, let alone bipedal apes gawking at mov-

* In his own time, Haeckel was equally famous for promoting certain aspects of Charles Darwin's research and developing his own influential theories of evolution, some of which were flawed and are now defunct. He coined several scientific terms still in use today, including *ecology* and *phylogeny*. He was also an advocate for eugenics whose writing, some historians contend, contributed to Nazi and fascist ideology.

ing sidewalks and incandescent light bulbs in the early twentieth century. If plankton did not exist, Earth would have no complex life of any kind. Without the innumerable viruses, bacteria, single-celled organisms, and as-yet-unclassified mysteries that we call plankton, the ocean would be completely unrecognizable: not a vast ecosystem replete with unexplored habitats and undiscovered species of inconceivable wonder—not the presumed birthplace of life and the foundation of the biosphere—but an immense volume of lonely water, brimming only with the silence of all that might have been.

CHAPTER

5

THESE GREAT AQUATIC FORESTS

SANTA CATALINA ISLAND DOES NOT SEEM LIKE A PLACE WHERE ONE would find lush forests. Located about twenty-two miles off the coast of Southern California, Catalina has a Mediterranean climate with hot, dry summers and mild winters. There are some woodlands and scattered trees on the island's rocky terrain—oak, ironwood, cherry, imported palms and eucalyptus—but far more abundant are various drought-tolerant aromatic shrubs mingled with grasses and cacti: plants like manzanita, sage, toyon, lemonade berry, buckwheat, and prickly pear. Yet Catalina and her sister islands are also home to some of the most vigorous forests in the world. Though they have their own canopies, understories, and floors, these forests do not require any soil, nor do they contain any wood. They can stretch more than 120 feet, but you'll never see them on the island's slopes or horizon. To find them, you must step away from the hills and beaches, leave the land and sky behind, and plunge through a liquid looking glass into a parallel world.

Lorraine Sadler knows Catalina's underwater forests intimately. A sprightly, dark-haired woman who animates her speech with quick, fluttering gestures, Sadler has been snorkeling and scuba diving around the Channel Islands for more than thirty years. After college, she worked for a marine laboratory in Marina del Rey that provided neuroscientist Eric Kandel with the speckled sea slugs whose unusually large and accessible neurons were essential to his Nobel Prize–winning research on learning and memory. In the late 1980s, Sadler moved to Catalina, where she became a scuba diving instructor and a hyperbaric chamber technician. She is a founding member of the Women's Scuba Association, a member of the Women Divers Hall of Fame, and a long-time marine science educator whose students praise her unremitting passion for the ocean and its inhabitants.

On a tranquil midsummer morning, I met Sadler in Two Harbors, an unincorporated community with about two hundred permanent residents and a single general store, situated on an isthmus in the north-west region of Catalina. We drove along narrow dirt roads toward a cove known as Howland's Landing, where we pulled on our wetsuits and snorkel masks, secured our weight belts, and carried our fins down a steep footpath to a beach of gray and pink pebbles. It had been several years since I had swum in the ocean, and it took me a little while to acclimatize. As I trod the cold water and struggled to adjust the straps on my mask, I accidentally dropped my underwater notebook—a bit of gear that I was excited to try for the first time. Sadler dove down with ease and retrieved it for me. "Thank you so much," I said, feeling as graceful in the water as a potato.

We started snorkeling in earnest, staying close to the edge of the ridge. With my eyes fixed downward, I began to orient myself to the world below—a community as lively as any tropical reef I'd ever seen. Rather than a prismatic citadel of coral, the rocky reef beneath us was plush with seaweeds and seagrasses in vivid shades of emerald, sepia, and chartreuse. The profusion and variety of this submerged vegetation reminded me of early spring hikes in protected regions of the Pacific Northwest, where ferns, sorrels, and brambles grow in thickets and trees drip as much with moss and lichen as with rain. Garibaldi damsel-fish, bright as tangerines, performed their looping courtship dances and tended their red algae nests, packed with thousands of eggs. Schools of silvery topsmelt flitted past us near the surface as camouflaged kelpfish, slender and striped, rocked back and forth with the currents. At one point we encountered an octopus seeping along the seafloor while flaw-lessly mimicking the color and texture of the surrounding rocks—a mesmerizing daytime sighting of a typically elusive nocturnal creature.

As exciting as these encounters were, I initially worried that Catalina's underwater forests were not as impressive as I'd heard. Most of the vegetation we'd seen so far, while beautiful, was fairly small. "Let's get a bit further out and see what's there," Sadler said, guiding us away from the ridge and toward the open sea. As we swam into deeper water, I started to notice squat bunches of kelp growing like kale among the rocks, interspersed with lone strands reaching for the sun. A couple minutes later, with surprising abruptness, we were completely surrounded by thick braids of kelp stretching from the ocean floor to the windswept surface, each one as imposing as a fairytale beanstalk. This is what we had come to see: giant kelp (*Macrocystis pyrifera*), the largest of all seaweeds and one of the fastest-growing photosynthetic organisms on the planet. In ideal conditions, giant kelp can grow more than two feet each day. "In one cove, we've clocked it at three feet a day," Sadler told me.

With a little help from our weight belts, Sadler and I dove into the forest, exploring every level. At the edge of the forest, where we could still see the seafloor, Sadler explained how the kelp clung to rocks with rootlike structures called holdfasts. Numerous air bladders helped their stems, technically known as stipes, stretch toward the sun. In the water, the kelp's puckered "leaves"—formally termed *blades* or *fronds*—ranged in hue from olive to mustard; at the surface, where they spilled upon each other in great heaps, they appeared much darker, at times almost chocolatey. The youngest and smallest blades were paper-thin to the touch, while the larger ones had a texture somewhere between leather and rubber. As the fronds billowed in the currents, they briefly revealed mottled kelp bass and blue perch, likely hiding from predators. Around Catalina alone, more than 150 fish species depend on kelp for habitat and food. Sea lions and harbor seals often hunt in kelp beds. Many birds and mammals, including whales, shelter themselves and their young in kelp during storms. Otters even use kelp as a kind of toddler leash, wrapping their pups in seaweed to prevent them from drifting away while they track down a meal.

Sadler combed through the vegetation, pointing out the small snails, worms, crustaceans, and other invertebrates that live on every part of the kelp, from its gnarled holdfasts to its floating mats. "This is so cool," she said. "Just look at all the species on this one blade. It's its own habitat—its own ecosystem." At Sadler's encouragement, I tried lifting some of the kelp out of the water, but the mere handful was so shockingly heavy that I could raise it above my head for only a few moments. "There's something else you've got to try," Sadler said. "It's called kelp crawling. A good way to move across the canopy." She demonstrated, pulling herself along the tangled fronds at the ocean's surface with a speed and agility that attested years of experience. As I tried my best to imitate her, I felt like nothing so much as a pea rolling around a bowl of creamy pappardelle.

Before this snorkeling trip, my interactions with living seaweed had been limited. Growing up in Northern California, where my family frequented the coast, my brothers and I sometimes integrated seaweed into our beach games, using it to decorate sandcastles, popping air bladders like bubble wrap, and leaping over pungent bundles of decaying wrack. While exploring tidepools, I undoubtedly overlooked many small seaweeds, preoccupied instead with the scarlet flash of a scuttling crab or the electric recoil of a sea anemone. As a teenager, I once kayaked across masses of kelp in Monterey Bay, but what I remember most clearly are sea otters cracking clams against rocks balanced on their bellies. Nothing had prepared me for the thrill of witnessing giant kelp in its native habitat. Only after I was ensconced in thousands of undulating, underwater leaves—only after weaving in and out of the enormous, golden-green stalks from which those leaves sprang and climbing atop their jumbled profusions—did I truly understand why we call them kelp *forests*.

KELP IS BUT ONE type of seaweed, which is itself a subset of marine algae. Like *plankton* and *microbe, algae* is one of those biological catchalls for a large group of organisms that have a lot in common despite having evolved on distant branches of the tree of life. *Algae* refers to more than fifty thousand photosynthetic species that range from the microscopic and single-celled (diatoms, coccolithophores, and dinoflagellates) to massive multicellular entities (giant kelp and bull kelp) and live in habitats as diverse as rivers, icebergs, tree bark, and sloth fur. Although kelp and other seaweeds look and behave a lot like plants, not everyone agrees that they are.* Compared to most land plants, algae have simpler anatomy: they don't have true roots that seek out and sponge up water and nutrients; instead, they absorb what they need directly into their cells. They generally lack the complex internal plumbing plants use to transport fluids through their bodies. And they do not flower or produce seeds.

Kelp forests, which favor rocky shores and cool, nutrient-rich water, grow extensively along 25 percent of the world's coastlines, flanking every continent, including Antarctica. Seaweeds as a whole are even more prevalent, inhabiting both temperate and tropical regions of the globe. The planet's oceans and coasts are further populated by many other types of aquatic vegetation: vast mangroves, salt marshes, and seagrass meadows, to name a few, some of which may be hundreds of thousands of years old.†

Although scientists have studied marine vegetation for centuries, only relatively recently have they developed the necessary tools to demonstrate and quantify the importance of these organisms in regulating global climate and calibrating ocean chemistry. Like terrestrial life, ocean-dwelling plants and macroalgae benefited from the ecological transformations wrought by the microbes that preceded them. In

* Scientists differ on how best to classify algae, plants, and other photosynthetic organisms. Most experts recognize three main groups of algae, each of which contains numerous seaweed species: the usually shallow-dwelling green algae, such as sea lettuce and water silk; the often feathery red algae, including the species that are pressed and dried into sheets of *nori* and used to wrap sushi; and the sometimes massive brown algae, the group to which kelp and sargassum belong. In the strictest proposed classification, terrestrial plants are the only true plants. A more inclusive system expands the Plant Kingdom to include green algae, the evolutionary ancestors of land plants. The most generous framework extends the moniker of *plant* to most marine algae, including many seaweeds. Whether seaweed qualifies as a plant, then, is a surprisingly personal choice.

† Seagrasses, which are indisputably plants, have an evolutionary history that parallels whales and dolphins: they evolved from land plants that returned to the sea about 100 million years ago, changing their cell structure and losing their leaf pores but keeping their roots, internal plumbing, and flowers, which continue to be pollinated by tiny crustaceans and marine worms.

many cases, marine vegetation made the ocean even more habitable, ultimately supporting a much more complex and diverse community of living creatures. One of the most remarkable demonstrations of this power may have occurred 50 million years ago in a hypothesized scenario known as the Azolla event. At the time, Earth was a steamy hothouse with crocodiles, turtles, and palm trees populating the Arctic. Deep-sea sediment cores indicate that a small but especially vigorous aquatic fern known as *Azolla,* which could double its biomass in less than two days, repeatedly formed thick mats across the Arctic Ocean. As the mats photosynthesized, they pulled huge volumes of carbon from the atmosphere, much of which was sequestered in the seafloor through the sinking and burial of dead vegetation. Some scientists have proposed that, over a period of eight hundred thousand years, *Azolla* shifted so much carbon from the air to the deep sea that it helped push Earth out of its former hothouse state and into a more familiar climate in which extensive glaciers and sea ice could envelop the poles.

In modern oceans, kelp forests create unique underwater climates and habitats by altering the distribution of sunlight beneath the sea's surface, the speed and direction of currents, and the rate at which marine snow sinks through the water column. Within a kelp forest, certain currents flow up to ten times slower, and eddies may be 25 to 50 percent weaker compared to nearby areas without kelp. Like coral reefs, mangroves, and marshes, kelp forests shield coastal communities from the brunt of storms, reducing wave heights by up to 60 percent. The comparatively calm waters of these underwater forests are ideal nurseries for the spores, eggs, and larvae of many creatures.

Seaweed can even reshape the ocean floor and relocate some of its most obdurate inhabitants. When turbulent currents and strong tides rip seaweeds from their moorings, their rootlike holdfasts sometimes take pieces of bedrock with them. If the detached seaweeds retain enough air bladders to remain highly buoyant, they can hoist rocks—even large boulders—up through the water column, like a bunch of weather balloons lifting an elephant. Seaweeds have even been observed levitating living mussels, clams, oysters, and scallops. After traveling anywhere from a few feet to thousands of miles, these seaweeds may wash up on the shore or sink to the bottom of the ocean, along with their payloads of rock, shell, and sediment.

Although scientists have long recognized terrestrial forests as a key component of global carbon cycling, the importance of marine vegetation has, until recently, been undervalued. Researchers have now demonstrated that marshes, mangroves, and seagrass meadows store carbon for decades to millennia in woody tissues, sprawling underground root systems, and layers of root-stabilized sediments that can be more than thirty-six feet thick. How rootless, soft-bodied, palatable seaweed might sequester carbon on long timescales has been less clear. Many scientists presumed that the vast majority of seaweed was eaten or rapidly decomposed, thus releasing its carbon back to the ocean and atmosphere. Even scientists who study carbon stored in marine and coastal ecosystems, known as blue carbon, have traditionally discounted seaweed as a meaningful carbon sink. The latest evidence, however, suggests that seaweeds and other algae have been a major component of the global carbon cycle for at least 500 million years and perhaps as long as two billion years.

Carlos Duarte, a marine ecologist at King Abdullah University of Science and Technology in Thuwal, Saudi Arabia, and his colleagues argue that seaweeds constitute the most extensive and productive form of coastal vegetation and that they are a massively undervalued carbon sink. Although many types of seaweed grow in rocky areas near the coast, they do not always stay there. Wind and waves can propel scraps of kelp and rafts of seaweed as large as Chicago (about 230 square miles) far from their origins, where they eventually break down into tiny particles. Global ocean surveys, Duarte notes, have found a "ubiquitous presence" of seaweed DNA up to 3,020 miles away from the coasts. Bits and pieces of seaweed have also been recorded sinking to a depth of four miles and have even been recovered from the guts of abyssal isopods—giant, deep-dwelling cousins of roly-polies. Strong currents racing along submarine canyons routinely carry large amounts of seaweed to the seafloor: by one estimation, the canyon adjacent to Monterey Peninsula alone conveys 130,000 tons of kelp to the deep sea each year. Storms in the North Atlantic may transport up to seven billion tons of kelp annually to the 5,900-foot-deep ocean floor off the Bahamian shelf. Once seaweed sinks below 5,000 feet or so, out of the reach of most predators and decomposers, its carbon is sequestered on "close to permanent timescales."

Like trees and other land plants, seaweeds continually secrete a variety of carbon-rich compounds, including sugars, terpenes, and dimethyl sulfide; it's not always clear why, but some of these molecules are likely involved in chemical signaling, immune defenses, and the cultivation of rich microbiomes. A portion of these exuded sugars and other compounds are eventually buried in seafloor sediments, along with shreds of seaweed. Some seaweeds are extremely resistant to decomposition. Bladderwrack, for example, stores up to a quarter of its dry mass in resilient sugars that only highly specialized bacteria can consume, using "one of the most complicated biochemical degradation pathways for natural material that we know of," as biologist Jan-Hendrik Hehemann once described it.

Given these revelations, many marine ecologists now think that seaweed sequesters far more carbon than was previously realized. Calculations suggest that every year nearly 610 million metric tons of seaweed-derived carbon dioxide settle in coastal sediments and the deep sea. Collectively, mangroves, salt marshes, seagrass meadows, and kelp forests have the capacity to store twenty times more carbon per square meter than terrestrial forests and may already sequester as much as 3.1 billion metric tons of carbon dioxide annually, about a third of the ocean's carbon uptake. An expert panel convened by the Energy Futures Initiative concluded that the cultivation of kelp and other seaweeds has the potential to draw down more than five billion tons of atmospheric CO_2 every year.

These insights are surfacing at a time when the threat of human activity to many types of marine vegetation is at its peak. By the early twenty-first century, the world had lost close to 30 percent of its seagrass meadows and somewhere between a third and a half of its mangroves. Kelp beds are highly dynamic ecosystems that grow and shrink dramatically from one season to the next and often recover rapidly following catastrophe. On average, however, global kelp forests have declined in the past fifty years due to a variety of stressors, including global warming, more frequent storms, overfishing, and pollution. In some places, such as Tasmania and Northern California, shifting currents, as well as the loss of otters and other important predators, have allowed legions of sea urchins to devour entire kelp beds, replacing them with crusty barrens that reduce water quality and impede nutrient cycles.

As the ecological understanding of marine vegetation deepens, scientists and aquaculturists around the world are increasingly interested in whether a combination of conservation, restoration, and farming can help preserve the biodiversity of the ocean's forests and grasslands and simultaneously mitigate some aspects of the climate crisis. In part because they are already widely grown for food and medicine and because they are so fast-growing, seaweeds have become the focus of a variety of climate-oriented ventures, including ocean farms, carbon-capture startups, and restoration projects involving lab-bred superkelps. Although many experts are genuinely excited about these prospects, some worry that seaweed will become yet another scapegoat for the fossil fuel industry and another false savior for climate activists. The ecological power of the ocean's forests is undeniable, but humans have a long history of trying to harness and subjugate other species, rather than working with them—often with severe repercussions. Breaking that cycle has never been more urgent.

WHEN MARTY ODLIN WAS a freshman in high school, his history teacher, Ms. Lee, asked the class a question: What will be the most significant event of your lives? "The fall of the Soviet Union," Odlin suggested. "Wrong!" Ms. Lee said. "It's going to be combating climate change." As Odlin remembers it, Ms. Lee was particularly interested in the importance of preserving and restoring forests as a way to sequester carbon. Odlin, who grew up in a multigenerational fishing family in Maine, spent the following summer as he did much of his youth: working on boats. While catching lobsters just outside of Casco Bay, Odlin's crewmates dredged up some kelp. One of the boatmen mentioned that kelp can grow several feet a day and form underwater forests. Odlin was astonished. "I remember spinning on that for a while," he told me, eventually connecting it to what Ms. Lee had said. As he grew older, he started to wonder if seaweed might also have a role in managing the climate crisis.

In college, Odlin studied art, architecture, and mechanical engineering. Later, he worked in product design and manufacturing and served as assistant director of the Education Center for Sustainable Engineering at Columbia University. In 2011, he returned to Maine to help his parents manage their fishing fleet. For the next six years, Odlin immersed himself in the day-to-day operations of a commercial fishery, fascinated by its inner workings. In parallel, he studied the history, economics, and ecology of fishing. A tour of fisheries in Iceland and Denmark impressed upon him the importance of sustainably harvesting the ocean's resources and the potential for technological innovation to help achieve that goal.

After Odlin's parents sold their fleet and retired, he contemplated getting a fishing boat of his own but eventually decided to try something completely different. In 2017, he founded Running Tide, an aquaculture company focused on developing new technologies to "leverage the ecological benefits of shellfish and seaweed farming." From the start, he knew that using kelp to sequester carbon would be one of his company's primary missions. He began working out how to do so in his backyard, "building prototypes out of flatwood, programming Arduino for the controls, and going to Radio Shack for parts." As of this writing, Running Tide has more than thirty employees and has received more than $15 million from investors such as venture capitalist Chris Sacca and customers like the e-commerce company Shopify and the Chan Zuckerberg Initiative.

Running Tide's headquarters in Portland, Maine, has the feel of a Silicon Valley hackerspace spliced onto a fishing pier warehouse. When I knocked on their doors one spring morning, a square-jawed, lightly freckled man named Adam Baske, then the company's head of business development, greeted me and introduced me to a few colleagues conversing by some computers and a whiteboard covered in diagrams of kelp's reproductive cycle. I followed Baske to a nearby loading dock, through a curtain of transparent plastic strips, and into a shack that housed several blue plastic tanks hooked up to an array of tubes and sensors. "This here is the biggest kelp hatchery in the country," he said, grinning. "Most kelp hatcheries in the U.S. are doing kelp in little 20-gallon aquariums. These are 250 gallons. Everything we do, we try to do it as big as possible." He lifted the lid from one of the tanks, revealing long fluorescent lamps attached to it with zip ties. Dozens of PVC pipes bobbed in the green water, each wound tightly with thin white rope.

Baske explained how he and his colleagues inoculate the rope with kelp spores and raise them in the hatchery for about thirty days before transferring the rope to buoys in the ocean, where the kelp can grow nearly fifteen feet in a few months. They've experimented with various species, focusing on an amber cousin of giant kelp called sugar kelp. "We want whatever is going to suck the most carbon up the fastest," he said. "That might mean different species for different regions of the ocean." Running Tide has also been developing new ways to efficiently farm huge numbers of oysters, using both onshore hatcheries and offshore growing systems that can be raised and lowered as needed throughout the year. In addition to being a highly desirable food with a small carbon footprint, oysters improve water quality and prevent harmful algal blooms by sponging up excess fertilizers and waste products that wash into the ocean.

A little later that day, Baske, Odlin, and I boated into Casco Bay, along with head of strategy Claire Fauquier, to see some of Running Tide's prototype kelp microfarms. Although it had been unseasonably warm the day before, the temperature had dropped significantly and storm clouds were stewing overhead. Odlin—mostly bald with a short beard and mustache; wearing a gray beanie, rain jacket, and shorts—huddled beneath some spare towels for extra warmth, periodically excusing himself to answer business calls. When we reached the shores of Cliff Island, Odlin leaned over the side of the boat and began to haul a yellow buoy out of the water, straining against its weight. As he pulled, a long bundle of rippling kelp emerged, like the tentacles of some enormous vegetal jellyfish. Although the blades were similar in color to those of giant kelp, they were much slimmer and more ruffled. Notably, they did not have any air bladders.

"That's what it looks like," Odlin said, turning to me with a broad smile. "Carbon-sucking machines. And it's just ever so slightly negatively buoyant, which is fantastic. There's like 400 pounds of kelp on that, and look at how little it draws on that buoy. You could pull in a lot of carbon and not add a lot of flotation, which is key to the model. Everything needs to be dematerialized."

Eventually, Running Tide plans to release thousands of kelp-laden buoys into the open ocean, relying on currents to guide them over extremely deep and flat regions of the sea known as abyssal plains. The buoys, which will be made from as yet undetermined biodegradable materials—possibly reclaimed waste wood and limestone—will gradually deteriorate, allowing the kelp to sink after three to nine months of growth. Once the kelp reaches a depth of 3,280 feet or more, its carbon should remain in the deep ocean for thousands of years—possibly eons. Using this system, Odlin and his colleagues ambitiously aim to sequester billions of tons of atmospheric carbon dioxide. Presumably, corporations will pay Running Tide for this service as a way to offset their past and present emissions. They hope to lower the cost to between $50 to $100 per ton.

Many climate experts have argued that although carbon capture ventures should never detract from the essential task of replacing fossil fuels with renewable energy, they will have a meaningful complementary role in managing the climate crisis. Several of the seaweed scientists I interviewed expressed cautious optimism about using kelp for carbon sequestration, quickly followed by strings of caveats and concerns. If Running Tide or similar companies, such as Pull to Refresh and Phykos, were to deploy legions of small buoyant kelp farms or solar-powered robot nurseries into the ocean, how would they keep track of them and confirm their fate? What if they collide with vessels, tangle propellers, endanger wildlife, sink too early, or fail to sink at all, ultimately becoming yet more polluting flotsam? Will kelp adapted to coastal environments actually grow to an appreciable size in the often nutrient-poor open ocean? If it does, will it outcompete vital plankton communities, disrupting fundamental ecological cycles in unpredictable ways? No one has ever attempted to deliberately grow and sink kelp on anything close to a scale that would make a difference to global climate. Even if these approaches successfully sequester atmospheric carbon, such a prolific influx of organic matter into mysterious deep ocean habitats will likely have unanticipated consequences.

"We don't want to greenwash what seaweed can do," says Nichole Price, a marine ecologist at Bigelow Laboratory for Ocean Sciences who has extensively studied seaweeds. "There are so many questions about where it will all go if someone sinks it. When it lands on the seafloor, does it create an anoxic area that is going to kill off a bunch of sea life? What is it going to get wrapped around on the way? It's exciting work, and it's certainly something that needs to be studied, but I would like to see the math that demonstrates it's going to be a net carbon sink, given all the inputs and costs. There are a lot of modeling exercises that would be really informative to do first."

Odlin is keenly aware of the many challenges and valid concerns that he and his colleagues must confront, and he says he is working closely with scientists to resolve them. Yet he remains as passionate about his venture as ever. "I'm an engineer and I'll caveat everything, but I know this works," he told me in one of our conversations. "I know that sinking kelp is a way to sequester carbon permanently. Every million tons of carbon is going to count. We have to chase every little bit down. The two big questions are: How big can we do this—at what scale? And what will the costs be? It's this kind of ironic thing where we're trying to save things that are priceless. You can't put a price on stopping global warming. It's an existential threat to humanity. There's no guarantee we can get to the point where the world can actually afford doing this at scale. But the world has to do something, right?"*

* I visited Running Tide's headquarters in the spring of 2021. In June 2024, by which point the first edition of this book had already been printed, Running Tide announced that it had begun to shut down its global operations due to a lack of sufficient funding. By fall of that year, however, Marty Odlin had reacquired the company's primary assets in hopes of continuing its core mission. For more details, see the author's note on page 240.

MINIATURE KELP FARMS DRIFTING through the open ocean may be novel, but the human affinity for seaweed is not. Archaeological evidence indicates that humans have been harvesting seaweed for food and medicine for at least fourteen thousand years. Some researchers have proposed that humans migrated from Asia to the Americas tens of thousands of years ago by canoeing along a resource-rich "kelp highway" stretching from the shores of modern-day Japan to Alaska and down the Pacific Coast to Chile. The Indigenous Peoples of the Pacific Coast tended beds of wild seaweed as though they were gardens. In the Pacific Northwest, native peoples ate raw young kelp; dried seaweed in cedar boxes; cured and twisted long strands of bull kelp into fishing lines, nets, and ropes; and fashioned hollow kelp parts into funnels, hoses, and storage containers. On both coasts, Indigenous Peoples lined cooking pits with seaweed for moisture and flavor before baking or steaming clams, lobsters, and other seafood.

Ancient documents and tablets record the harvest and consumption of seaweed in Asia since the fifth century, though the practices certainly began far earlier. The Taiho Code, established in Japan around A.D. 700, mandated a tax from the provinces to the central government, often in the form of valuable goods, including silk, lacquer, and seaweed. In the sixteenth century, Japanese fishermen discovered that seaweed readily grew on the bamboo pens they built to hold fish. Under orders from the reigning *shogun,* they began cultivating seaweed by throwing bundles of bamboo and camellia branches into estuaries, gradually refining their methods over the centuries. By the 1800s, the Japanese were using papermaking techniques to produce neat dried sheets of *nori* in huge quantities. Meanwhile, the Chinese had learned to turn red seaweed into an "iced jelly" that was sold as a refreshing summertime treat.

In ancient Greece, the philosopher and botanist Theophrastus described great seaweed forests in the deep Atlantic, in particular sugar kelp of "marvelous size" growing near the Pillars of Hercules. Pliny the Elder wrote of prescribing seaweed for gout and swollen ankles, the Greek poet Nicander said it was a cure for snakebite. Ancient Arab mariners were "familiar with many kinds of seaweed, using them as an aid to navigation, recognizing that seaweed could provide information on winds, tides, depth, and conditions on the sea bottom," as Kaori O'Connor writes in *Seaweed: A Global History.* According to some accounts, seaweed even saved Arab navies from the horrors of Greek fire, a weapon similar to a flamethrower that spewed a mysterious combustible substance that continued to burn on the surface of water. Allegedly, Abd al-Rahman, master of the Alexandria shipyards, coated warships with a fire retardant extracted from brown seaweed. The key compound, alginic acid, is still used to fireproof fabric today.

Many thousands of years ago, in Ireland, Wales, and Scotland, coastal hunter-gatherers collected seaweed, along with nuts, berries, and lichen. Iceland's oldest surviving law book, written in the twelfth century, discusses the legal right to collect a red seaweed known as dulse on another's property. Early farmers spread seaweed across their fields as fertilizer and fed it to their livestock. On some remote Scottish islands, a breed of sheep confined to the shoreline adapted to a diet composed almost entirely of seaweed—the only terrestrial animal to do so, apart from marine iguanas. By the seventeenth century, people in various parts of Europe were burning seaweed in vast kilns and using the ashes, which were rich in sodium and potassium, to make soap and glass.

Today, more than thirty million metric tons of seaweed are farmed around the world each year. Asia has many of the oldest, largest, and most intensive farms, with China, Indonesia, Korea, Japan, and the Philippines responsible for more than 97 percent of modern production. A single bay in China, the approximately fifty-square-mile Sanggou Bay, produces more than 240,000 tons of seafood each year, including 80,000 dry tons of kelp—about a third of the kelp produced in the entire country. As a food, seaweed is enjoyed in just about every imaginable way: raw and whole; chopped, sliced, and shredded; simmered, boiled, baked, and toasted; smoked, powdered, pickled, fermented, distilled, and jellified. Some seaweed connoisseurs talk about their favorite algae as though they were fine wines. Food writer Harold

McGee explains that, depending on its chemistry and how it is prepared, a seaweed might have a fishy, floral, or spicy aroma with notes of bacon, cooked corn, black tea, or hay. Compounds extracted from various seaweeds are widely used to thicken, emulsify, and stabilize foods as diverse as eggnog, lasagna, fish fingers, coleslaw, ketchup, and frozen cheesecake. Seaweed is also the source of agar, the gel that scientists use to cultivate microbes in petri dishes, as well as binding agents for medical tablets, wound dressings, and dental impression putties.

Although Asia still has the most robust seaweed aquaculture industry by far, it is an increasingly popular enterprise in other parts of the world. Between 2014 and 2016, the coastal area in Norway dedicated to seaweed farming tripled. Norwegian scientists have estimated that if the country continued to expand seaweed farms along its extensive coastline, it could harvest twenty million tons by 2050. In the Thimble Islands of Long Island Sound, Bren Smith has established one of America's first "3D" ocean farms, which uses a scaffolding of ropes, buoys, and anchors to grow many different species in a relatively compact area. Depending on the season, long vertical streamers of kelp, scallops, and mussels hang off ropes suspended horizontally near the surface, while oysters and clams reside beneath them on the seafloor. His twenty-acre farm sops up pollutants and excess nutrients while producing about 100 tons of sugar kelp and 250,000 individual shellfish every year—all without the need for arable land, fresh water, or fertilizers.

At present, the world's seaweed farms can capture only 2.5 million tons of CO_2 annually, a tiny fraction of the more than 36 *billion* tons of carbon dioxide humans release into the atmosphere each year. But seaweed farms currently occupy a mere six hundred square miles globally, which is only 0.04 percent of the area populated by wild seaweed, which, due to pollution, warming seas, ravenous urchins, and ecological cascades triggered by human activity, is itself much smaller than the seaweed populations of past centuries. Antoine De Ramon N'Yeurt, a marine botanist at the University of the South Pacific, and his colleagues estimate that if seaweed farms covered 9 percent of the world's oceans, they could capture at least 19 billion tons of carbon dioxide each year, more than half of global emissions. Nine percent of the world's oceans is a massive area, about twice the size of Russia—a wildly ambitious goal. But these extreme figures underscore seaweed's immense potential. If sinking kelp to sequester carbon eventually proves too costly, environmentally damaging, or otherwise unfeasible, there are still many other ways to amplify seaweed's ecological benefits. Even a moderate expansion of seaweed farms throughout the world, combined with the restoration of native seaweed communities, could make a major contribution to the drawdown of carbon emissions.

Of course, if every scrap of cultivated seaweed were harvested and consumed, then all the carbon within them would quickly return to the atmosphere. But as recent research has shown, wherever wild and cultivated seaweeds grow, their continually deteriorating tissues and exuded sugars often sink into deep water and settle on the seafloor, sequestering far more carbon on long timescales than was previously realized. Seaweed farms may even enhance carbon capture by slowing currents and allowing more organic particles to sink, settle, and gradually form thick layers of sediment. Duarte, Price, and dozens of collaborators have launched an international research project to investigate exactly this: they are collecting sediment cores from seaweed farms in North America, Europe, and Asia—including a three-hundred-year-old farm in Japan and a massive farm in China that is visible from space—and measuring their carbon content, among other characteristics. Preliminary results are encouraging.

Beyond trapping carbon and improving water quality, seaweed and other forms of marine vegetation also minimize and sometimes even reverse ocean acidification, at least at a local scale. Several studies in the past few years demonstrate that as seaweed and seagrass absorb carbon dioxide from the water, they create high-pH refuges along the coast, sheltering crabs, oysters, mussels, and other farmed and wild species from acidifying waters that dissolve their shells. Much of this research has occurred on the Pacific Coast, where wind and ocean currents routinely bring deep and nutrient-rich, but highly acidic, water to the surface—water that absorbed carbon humans pumped into the atmosphere decades ago. When a team of collaborating scientists and aquaculturists cultivated sugar and bull kelp on a scaffolding of rope in open water northwest of Seattle, they observed significantly less damage to the shells of oysters, mussels, and snails inside the kelp farm compared to those in sites beyond its borders. Field studies in Oregon and Washington State have demonstrated that juvenile Pacific and Olympia oysters grow 20 percent faster, and are more likely to survive, when they live in beds of eelgrass. Researchers in California estimated that a restored kelp forest off Palos Verdes Peninsula temporarily raised local pH by as much as 0.4 points, corresponding to a decrease in acidity of up to 60 percent.

In recent years, some scientists, aquaculturists, and entrepreneurs have become increasingly excited about a surprising use of seaweed that may dramatically reduce greenhouse gas emissions from land-based agriculture. As microbes in the digestive tracts of cows, sheep, goats, and other ruminants break down plant tissues, they produce methane as a byproduct, which the animals belch in copious amounts. Although methane does not remain in the atmosphere nearly as long as carbon dioxide, it is about eighty times more effective at trapping heat in the first two decades after it is released. Globally, livestock contribute somewhere around 15 percent of humanity's greenhouse gas emissions. A modest but growing body of research has demonstrated that feeding cattle small amounts of seaweed—namely, feathery red seaweeds in the *Asparagopsis* genus—reduces their methane emissions by up to 80 percent without altering the taste of their milk or meat. Such seaweeds contain compounds, such as bromoform, that appear to inhibit gut microbes from producing methane.

As with using kelp for carbon sequestration, however, enthusiasm could easily leapfrog evidence. *Asparagopsis* is tricky to grow, and no one has figured out how to farm it at scale, let alone produce sufficient quantities for the planet's nearly 1.5 billion cattle. The long-term effects of giving livestock seaweed—including potential toxicity or birth defects—are not yet known, nor is it clear how such exposure would steer the evolution of the microbes in their guts. Ongoing research will ultimately help determine whether cattle-fed seaweed supplements meaningfully restrict methane emissions or remain as ineffectual as a garnish.

THE AFTERNOON AFTER WE snorkeled in Catalina's kelp forests, Lorraine Sadler and I drove to the opposite side of the island to visit a cove called Little Harbor. The cove was small and ordinary, one of the Pacific's many parentheses. A dozen or so people were enjoying the pleasant weather, kayaking, playing paddleball, and lounging in folding chairs.

I was not entirely sure why Sadler had brought us here. We were not planning to snorkel along this part of the coast. She had mentioned, however, that she often found large quantities of beached seaweed in the area. A brown curtain of kelp fringed the length of the shore, swaying in the surf. The sand was strewn with all manner of dismembered algal anatomy: limp kelp blades, ruptured air bladders, tangles of blackened wrack, and heaps of what looked uncannily like giant ramen and soba noodles. Sadler, wearing a shirt festooned with colorful fish and turquoise sneakers with neon pink laces, eagerly explored the beach, eyes fixed downward, frequently stooping to inspect a clump of seaweed or something half-buried in the sand.

At one point, she spotted an essentially intact giant kelp tumbling in the waves—holdfast, stipe, blades, and all. A storm had likely ripped it from the seafloor. As the ocean retreated, we both took hold of the stipe and dragged the entire kelp further up the shore. Sadler knelt and began to explore the holdfast, which was about the size of a watermelon but shaped like a partially flattened funnel. The exterior looked like a mat of golden-brown roots, dense and knotty, as though they'd been potbound for too long, whereas the concave underside that had once clasped a rock resembled a large bird's nest.

Sadler, ecstatic about the opportunity to study a prime specimen up close, began to pry apart the interlaced branches in the holdfast's underside. "This is amazing!" she said. "There's always so much to see."

Kneeling beside her and peering closer, I started to realize that the kelp was much more than kelp. Every available surface, Sadler explained, was populated by invisible microbes and rimed with white crusts formed by tiny aquatic invertebrates known as bryozoans or moss animals. Every crevice was stuffed with life or the signature of the living—a feather; a shell; another, smaller seaweed. Sadler continued to dissect the holdfast, naming what she uncovered. There was a delicate frilled red alga, like a feather from some miniature songbird. The ghostly casings of calcareous tube worms. Juvenile sea stars. Snails, shrimp, and tiny burrowing clams.

When Charles Darwin encountered kelp forests around Tierra del Fuego in 1834, he was captivated by the diverse assemblage of life within them: "The number of living creatures of all Orders, whose existence intimately depends on the kelp, is wonderful," he wrote in his diary. "On shaking the great entangled roots, a pile of small fish, shells, cuttle-fish, crabs of all orders, sea-eggs, star-fish . . . and crawling nereidous animals of a multitude of forms, all fall out together. . . . I can only compare these great aquatic forests of the southern hemisphere with the terrestrial ones in the intertropical regions."

The more I thought about those "great entangled roots," the more meaningful they became. Here was a jungle within a jungle, formed by an organism whose very presence made the ocean more habitable and whose growth or demise determined the fate of coastal ecosystems around the world. This was the mooring of a life form on which humans have depended for tens of thousands of years and whose power our species is now trying to exploit in new ways. The holdfast Sadler and I examined on the beach was, much like our planet, convoluted, fecund, and mysterious—seemingly infinite in its hidden complexity. The closer we looked, the more there was to see.

Time and again in the Anthropocene, we find that we have blundered into the same tragic predicament: through increasingly sophisticated science, we are finally deciphering some of the planetary rhythms that life and environment have coevolved over great spans of time, just as our widespread destruction of Earth's ecosystems and reckless consumption of fossil fuels threaten to distort or extinguish those very rhythms. We are rapidly gaining new appreciation for the many ways that life tends to stabilize and regulate Earth, while finally reckoning with the fact that our species has too often done exactly the opposite, pushing the planet into a state of crisis. Scrambling for solutions, we find that we know enough to recognize and even quantify the importance of the astoundingly complex ecosystems we inhabit but not always enough to confidently intervene when they begin to collapse.

Yet the sheer complexity and staggering diversity of our living planet are also reasons for hope, courage, and perseverance, because it is precisely this intricacy that makes Earth so resilient. As the geological record reveals, the world's ecosystems are replete with possibility, even when they're on the precipice of obliteration. If our species finally learns to work with Earth's ecosystems, as *part* of them, instead of trying to subdue them—if we address the source of the current crisis by fundamentally changing our relationship with the planet, rather than clinging to industrial and economic systems that were never sustainable—we will avert total calamity in the decades to come, minimize suffering, and ultimately create a better world. It won't be exactly like the Earth we've known, but it will be a world where spring is still full of song, snowmelt still feeds mountain streams, and forests still soar through the sea.

CHAPTER

6

PLASTIC PLANET

KAMILO BEACH HAS ALWAYS BEEN A PLACE WHERE THE OCEAN HOARDS its wrecks and wonders. A confluence of trade winds, currents, and geography concentrates flotsam on this remote and undeveloped crescent of rock and sand near the southeastern tip of the island of Hawaiʻi. When Indigenous Hawaiians searched for wood to hew into canoes, they sometimes visited Kamilo, where they would find giant logs that had drifted to the island all the way from the coniferous forests of the Pacific Northwest. The bodies of those lost to the sea often washed up there. And in some folktales, people used the dependable currents surrounding Kamilo to deliver messages to their loved ones.

In more recent history, Kamilo became infamous for accumulating appalling masses of a material that did not exist in any appreciable quantity a century ago but is now ubiquitous in the Earth system. Between the 1970s and the turn of the century, beachcombers, campers, and other visitors to Kamilo encountered mats of plastic waste completely covering the sand and heaps of plastic allegedly reaching eight to ten feet high. Some media outlets dubbed Kamilo "Plastic Beach" and described it as one of the "dirtiest" shorelines in the world.

In the mid-2000s, Charles Moore, an experienced sailor and environmentalist, had the opportunity to see the mounds of trash on Kamilo for himself. Several years earlier, he had started writing research papers and articles about plastic pollution in the Pacific Ocean—a subject that was quickly becoming the focus of his life and work. In 1997, while sailing from Hawaii to Southern California, Moore traveled through the North Pacific Gyre, an enormous clockwise vortex of ocean currents surrounding the Hawaiian archipelago. For days at a time, he never saw another vessel. "Yet as I gazed from the deck at the surface of what ought to have been a pristine ocean, I was confronted, as far as the eye could see, with the sight of plastic," Moore later wrote. "It seemed unbelievable, but I never found a clear spot . . . no matter what time of day I looked, plastic debris was floating everywhere."

Moore had sailed into the midst of what eventually became known as the Great Pacific Garbage Patch, one of at least two concentrations of buoyant debris within the North Pacific Gyre. Although popularly called "patches," they are not so much cohesive islands of trash as whirlpools pervaded with diffuse plastic confetti, clumps of fishing gear, and other plastic waste. The size of the Great Pacific Garbage Patch is indeterminate, but researchers estimate that it spans about 620,000 square miles—more than three times the area of Spain—and contains 1.8 trillion pieces of plastic, which works out to more than two hundred pieces for every person on Earth. In a landmark study, Moore and his colleagues determined that within the patch, plastic outweighs plankton six to one.

To get to Kamilo, Moore endured "a jarring off-road hourlong trek over barely there trails pocked with dust pits and sharp lava," as he described it in his 2011 book *Plastic Ocean*. The beach boasted "the attributes of a world-class tourist destination—mist-shrouded mountain backdrop, crescent bay, tide pools carved from lava, murmuring surf, and what appear to be sandy beaches," Moore wrote. But it was also "a literal dump." Surveying the beach, he encountered "plastic spray nozzles, product bottles, shoe parts, Nestlé coffee lids, toothbrushes, butane lighters," and fishing nets by the ton. Tiny plastic fragments were everywhere—not just on the sand but deep within it, too. Digging around, he found many "shiny little spheres," which he instantly recognized as nurdles: plastic pellets melted and molded into all kinds of commercial products.

Other materials on the beach were startlingly unfamiliar to him—neither purely geological nor entirely artificial. They appeared to be strange chimeras of rock and plastic, perhaps fused by the heat of lava flows. One looked like a gray plastic buoy that had melted onto basalt. Another strikingly colorful example seemed to consist, at least in part, of half-dissolved fishing nets stuck together. Several years later, during a public talk at Western University in Ontario, Moore presented photos of the unidentified amalgams and mentioned that he was looking for a geologist to visit Kamilo and study them up close. Patricia Corcoran, the Earth sciences professor who had arranged the talk, leapt at the opportunity. Kelly Jazvac, an artist who had been in the audience and was interested in environmental pollution, offered to assist.

In the summer of 2013, Corcoran and Jazvac, with the guidance of knowledgeable locals, drove a jeep down the same dirt-and-lava-rock road Moore had used to reach Kamilo. As soon as they stepped out of the vehicle and onto the beach, they found examples of the conglomerates Moore had documented. Some were as small as a grape, others as wide as a microwave. Some looked like lava rocks whose nooks and crannies had been stuffed with wads of gum or dribbled with candle wax. Others were chaotic jumbles of wood, stone, shell, coral, and plastic, like hunks of refuse regurgitated by a malfunctioning trash compactor. A few were smoothed and rounded, indicating that they had been repeatedly tumbled by tide and wave. In one area, they found molten plastic adhered to rock and sand nearly six inches below the surface.

Corcoran and Jazvac searched for evidence to support Moore's hunch that volcanic activity had fused the different materials, but they soon learned that there had not been any molten lava in the area for more than a century. From their own observations, however, and through conversations with local residents, they realized that people sometimes made campfires on Kamilo. The beach was so thoroughly steeped in plastic that it was impossible to find a spot where the heat of a fire would not melt plastic debris. The melted plastic had become a matrix binding the geological, the biological, and the technological. By making fires in an area with such diverse substrates—a part of the planet where land and sea merged, where the ocean mingled the human and nonhuman, the animate and inanimate—our species had inadvertently forged a material that had never existed before. The odd conglomerates that Corcoran had collected were essentially a new type of rock—and she was the first geologist to examine it up close. In a 2014 paper in the Geological Society of America's monthly magazine, Corcoran, Jazvac, and Moore formally proposed a name for their discovery: *plastiglomerate*, the first rock type in the history of Earth composed partially of plastic.*

FROM THE TIME EARLY humans began carving fishhooks out of bone to the first wind-powered transatlantic voyages to the modern era of colossal cruise ships and robotic submersibles, our species has altered the ocean in diverse and enduring ways. We have devastated the planet's tropical reefs and pushed numerous populations of sea creatures to the brink of collapse in pursuit of their flesh, fur, oil, and blood. We have allowed excess fertilizers to seep into the ocean, seeding toxic plankton blooms, and fouled coasts and open water with immense volumes of oil. We have overwhelmed the ocean's soundscapes with a cacophony of sonar, seismic surveys, and marine traffic. We have opened new conduits between oceans, excavated submarine tunnels, crisscrossed the seafloor with communication cables, and experimented with new methods of extracting precious metals from vulnerable deep-sea ecosystems. And we have made the ocean warmer and more acidic than it's been in millions of years.

The staggering influx of plastic pollution to the ocean exemplifies the unparalleled speed with which our species has transformed so many different parts of the planet all at once. Throughout history, life has repeatedly introduced novel substances to the Earth system, some of which—such as highly reactive free oxygen (O_2) and indigestible plant tissues like lignin—were initially problematic or even lethal for many

* In conversation, Corcoran typically prefers to describe plastiglomerate as a stone rather than a rock, because the formal definition of "rock" is a mineral aggregate produced exclusively through geological processes without human involvement. Yet some researchers, such as geologist James Underwood, have proposed that the three main types of rock—igneous, metamorphic, and sedimentary—should be joined by a fourth: anthropic rock.

species. But these introductions typically spanned millennia or longer, granting ecosystems considerable time and opportunity to adapt. In contrast, we have flooded the planet with plastic in a geological instant, the repercussions of which we are only beginning to understand. If plankton are atoms of ocean, defining the planet's liquid chemistry, and seaweed is the fabric of Earth's great aquatic forests, forming vast underwater habitats and sheltering coastal communities, then plastic pollution is an insidious corruption of both, literally made from the remains of ancient plankton and algae but persistently undoing their ecological labor in ways both conspicuous and obscure.

The specific materials most familiar to us as "plastic" are relatively recent inventions, but the larger class of materials to which they belong—polymers—is ancient. A polymer, from the Greek for "many parts," is a substance composed of giant molecules that are themselves made of numerous repeating molecular subunits chained together. Polymers are abundant in nature: examples include DNA, muscle tissue, hair and nails, silk, cotton, wool, and many other fibers and resins produced by plants and animals as well as highly viscous forms of petroleum such as bitumen.

Humans began using polymers long before recorded history. At least seventy thousand years ago, humans were collecting bitumen where it seeped to the surface of the earth and using it both ornamentally and pragmatically. Over time, people learned how to use bitumen to affix handles to flint tools and to waterproof baskets, jars, roofs, and boats. Between 40,000 and 55,000 years ago, humans likewise used pine resin and birch bark tar as glue. Later, a full-fledged industry developed around conifer resin in ancient Rome. Some four thousand years ago, Mesoamericans invented what is arguably the first partly synthetic plastic. By combining the milky white latex of certain trees with the juice of morning glory vines, they created rubber, which they formed into sandals, hafting bands, and balls used in both games and rituals. When Europeans encountered rubber for the first time in the Americas, they had never seen anything quite like it: a material simultaneously sturdy and elastic—a solid that could bounce.

For millennia, people used ivory, horn, tortoiseshell, and other animal polymers to make everything from combs, buttons, and cutlery to piano keys and billiard balls. But animal polymers were not always ideal for their assigned roles or amenable to mass production. In the early 1900s, Belgian chemist Leo Baekeland began searching for a synthetic substitute for shellac, a resin extracted from insects through a slow and laborious process and valued as an electrical insulator. By combining certain proportions of phenol and formaldehyde under pressure, he produced a lightweight yet resilient material that was an excellent insulator and held its shape once molded, even when subjected to subsequent heat. He named it Bakelite and helped popularize use of the word *plastic* from the Greek *plastikos,* meaning moldable. Soon, Bakelite was mass-produced as components for phones, irons, toothbrushes, radios, cars, and washing machines. Not long after, researchers at the American chemical company DuPont invented neoprene, Teflon, and nylon. Nylon stockings became an international sensation. Stores sold out in hours, and shoppers brawled over limited supplies.

During World War II, annual plastic production in the United States nearly quadrupled from 96,600 metric tons in 1939 to 371,000 metric tons in 1945. The military used plastic to manufacture parts for airplanes, antennae, mortar fuses, and bazooka barrels. Nylon helped make parachutes, ropes, helmet liners, and body armor. Plexiglass formed aircraft windows. Teflon held volatile gases. Around the same time, innovations in injection molding facilitated the precise and efficient mass production of plastic. After the war, plastic's commercial applications proliferated. Cheap, versatile, lightweight, waterproof, and durable, plastic morphed into Tupperware, shopping bags, drinking bottles, packaging, and numerous other replacements or alternatives to many traditional uses of wood, paper, glass, and steel. Since then, global production has boomed. The world has produced a cumulative 8.3 billion metric tons of plastic since 1950. Annual global production is now about 360 million tons. More plastic has been manufactured in the past two decades than in the entire second half of the twentieth century.

In everyday conversation, we tend to lump all plastic together and speak of it in the singular, but the materials we're referring to are more formally known as plastics because they are so numerous and diverse. Today, there are hundreds of plastics with distinct chemical compositions and utilities. Some of the most widely produced and most familiar types are polyethylene and polypropylene, which are primarily used to make flexible films and similar materials for packaging, as well as car parts, pipes, and housewares. PVC and polyurethane typically find purpose in construction and in the automotive industry. Polyethylene terephthalate (PET) is favored for drink bottles and textiles. Polystyrene is often used for protective packaging and insulation, in both solid and foamed form. And polycarbonate usually becomes rigid transparent products like eyeglasses and greenhouses. The vast majority of modern plastics are made from oil and gas, which are first subjected to intense heat and pressure to reduce them to molecular building blocks rich in carbon and hydrogen, such as ethylene and propylene. These relatively small molecules are then chemically linked into new, much larger molecules, producing viscous resins that can in turn be processed into the powders and pellets used in injection molding. Thus, most of the plastics we use today are yet another form of fossil fuel.

Plastic pollutes the ocean through many different routes, including deliberate dumping and accidental littering at sea, but 80 percent of marine plastic debris originates on land. Every year, 8 to 12 million tons of plastic trash spill into the ocean, primarily through more than a thousand small- to medium-sized rivers in Asia, where dense populations use large amounts of disposable plastic but often do not have adequate waste-management systems—a problem exacerbated by a torrent of plastic from the United States and other wealthy nations attempting to outsource their own waste-management needs.* River networks close to the coast that accumulate large amounts of urban trash and receive heavy rainfall are especially likely to pollute the sea.

* For many years, the United States exported much of its plastic waste to China, where its fate was unclear. In 2018, China stopped accepting most imports of plastic trash. The estimated rate of plastic recycling in the United States subsequently dropped from an already low 9.5 percent to an abysmal 5 percent.

By some estimates, if current trends continue, humanity will have produced 33 billion cumulative tons of plastic by 2050, and the volume of plastic waste at risk of polluting the ocean each year will swell to 150 million tons, nearly twice the weight of fish caught at sea annually.

Most of the plastic that enters the ocean floats, at least at first. Sunlight, oxygen, and waves begin to degrade floating plastic debris, making it brittle. Certain microbes, fungi, algae, shellfish, and other sea creatures populate plastic waste, decreasing its buoyancy. As plastic in the ocean breaks apart and sinks, it is often consumed by larger organisms, such as fish and turtles. Other pieces of plastic may continually rise and fall as their buoyancy changes or repeatedly wash ashore with the tides. The plastic that remains in open water is thought to fragment into smaller and smaller pieces—known as microplastics and nanoplastics—and eventually settle on the seafloor, but there is still a lot of mystery surrounding its ultimate fate. Given how much plastic humans have produced in the past half century, there should be hundreds of millions of tons of plastic in the ocean, much of which should be floating on the surface. Yet surveys have found only a fraction of that amount atop the sea. At least some of that plastic might be buried in shorelines or at the bottom of the ocean or swallowed up by life—or maybe something happens to it of which we are completely unaware. As best as we can tell at present, the great majority of plastic flowing into the ocean is curiously missing.

Plastic has polluted far more than the seas. Scientists have found tiny plastic particles in just about every part of the Earth system: in rivers, lakes, and ponds; in rainforests, savannas, and mountain ranges; in polar ice and snow; in the soil, atmosphere, and rain; and in human lungs and blood. It is in and through the ocean, however, that plastic may leave its most enduring mark on the planet. The precise longevity of plastic objects polluting the sea and other parts of the planet is not yet known, but scientists think it might be in the range of hundreds to thousands of years. When plastic is buried deep underground or at the bottom of the ocean, it may survive far longer. Researchers have found plastic pollution in deep-sea sediment cores from the Mediterranean Sea and the North Atlantic and Indian oceans as well as more than three miles below the surface in the Kuril-Kamchatka Trench in the northwest Pacific, with concentrations as high as two hundred pieces of microplastic for every square foot. Much like the shells and skeletons of plankton that likewise accumulate in seafloor sediments, this plastic will eventually be compressed into rock and subsequently melted in the planet's molten interior or uplifted as layers in new mountains and cliffs.

When plastic is buried in sediment on land or at sea, it also has the potential to fossilize. Plastic made from fossil fuels is ultimately of biological origin. Many recalcitrant organic structures and residues have survived in the fossil record for thousands to millions of years: wood, spores, pollen grains, resins, and the intricate husks of plankton, for example. "It seems that many plastics will behave similarly over geological timescales," Patricia Corcoran, Jan Zalasiewicz, and their colleagues write in one paper. Plastic items, they explain, "may be fossilized in 'cast' and 'imprint' form even if all the original material is lost through biodegradation." Thus the outlines of ballpoint pens, plastic bottles, or compact discs "may be found as fossils in sedimentary rock in the future even if the plastic itself has degraded or been replaced by other materials." Other plastic objects may fossilize more or less like a dinosaur bone, with their 3D structure preserved.

Since Corcoran's first trip to Kamilo, researchers have formally documented several different types of plastiglomerate and similar hybrid materials on beaches throughout the world. Pyroplastics, for example, are characterized by an amorphous and dull-colored matrix of melted plastic. Some are so similar in color and texture to ordinary beach pebbles that they're all but impossible to distinguish by sight alone; only their unexpected lightness in the hand gives them away. When plastic fuses with rock, it becomes even more durable. Some scientists have proposed that plastic and plastiglomerate will become a significant part of the geological record—a unique signature of our moment in Earth history.

Because modern plastic is synthesized in laboratories and factories, it is often regarded as an "unnatural" material. Yet the concept of the unnatural makes sense only in contraposition to the idea of the natural, which is itself dependent on the false premise that humans and human artifacts are somehow separate from nature at large. In truth, humans are as much a part of nature as any other living creature. We are flesh and blood animals with bodies and behaviors shaped by evolution. We are not unique in possessing consciousness, culture, or communication. Our technologies are essentially far more elaborate versions of the spider's web, the bird's nest, and the monkey's stone hammer. And we are far from the only creatures that dramatically alter their local environments, create long-lasting infrastructure, and change the planet as a whole. The combined speed, magnitude, and diversity of the changes we have wrought is exceptional, but that is a difference of degree, not kind.

Everything our species fabricates is a modification of what nature has already provided. Plastic is yet another way of rearranging existing molecules. One could argue that modern synthetic plastics constitute molecular configurations that evolution would never have discovered on its own. Another way to look at it is that evolution discovered plastic through us. The trouble with plastic is not that it is unnatural, but that it is, like oxygen and lignin before it, entirely unfamiliar to the Earth system and its longstanding rhythms. The problem is that, in its current form, plastic is pervasive, highly resistant to degradation, and detrimental to many forms of life.

DARRELL BLATCHLEY KNOWS WHAT death by plastic looks like. He knows what it smells and feels like, too. An environmentalist and museum curator in Davao City in the Philippines, he routinely performs necropsies on marine mammals in order to determine cause of death and preserve their bones for educational purposes. Early one morning in March 2019, he received a call from the country's Bureau of Fisheries and Aquatic Resources about a sick whale in Davao Gulf. Local residents had observed the whale leaning heavily to one side and vomiting blood. Despite frantic efforts to save the animal, it had died by the time Blatchley arrived on the scene, floating on its side in the water, the shape of its ribs visible beneath a severely emaciated frame.

With help from onlookers, Blatchley and his colleagues hauled the whale onto a large trailer and towed it back to their museum, where they determined that the animal was a young male Cuvier's beaked whale, about fifteen feet long and 1,100 pounds. He was mottled gray and black with a slightly humped head and two underdeveloped tusks in his jaw. Whereas most members of his species were sleek, elongated teardrops, his body was unusually sunken in some areas and distended in others. His abdomen was so swollen and rigid that Blatchley had initially thought the whale might be a pregnant female.

As soon as Blatchley sliced open the whale's stomach, he was transfixed by the horror of its contents: the largest mass of compacted plastic waste he'd ever found in a single animal. He unfurled a tattered piece of a yellow plastic bag, perhaps from a banana plantation, followed by a black plastic trash bag, then another yellow one. Blatchley shook his head. The plastic just kept coming. Some of it had been compressed in the whale's stomach for so long that it had started to calcify, forming rock-hard chunks. "There were clumps I couldn't even pull apart," Blatchley recalls. "They looked like something that had been melted together." In total, he pulled eighty-eight pounds of plastic waste from the whale's body, including sixteen fifty-five-pound rice sacks, four banana plantation bags, numerous grocery bags, and tangles of nylon rope. All that trash comprised 8 percent of the whale's weight and completely blocked the passage from stomach to intestines, denying him water and sustenance. In some places, the whale's stomach acid, failing to digest the plastic, had eaten holes in the tissue lining the stomach instead.

Cuvier's beaked whales typically feast on squid and fish, relying on echolocation to find their prey. They can easily mistake undulating bags and other drifting plastic for food. The more plastic they consume, the weaker they become; deprived of the energy required to dive into deep waters, they are forced to feed closer to the surface, where they are more likely to encounter large pieces of debris. As of late 2022, Blatchley has necropsied seventy-five whales and dolphins. He estimates that plastics killed fifty-five of them.

The former Homestake Mine in Lead, South Dakota, now the site of the Sanford Underground Research Facility, the deepest subterranean lab in the United States. JAMES ST. JOHN

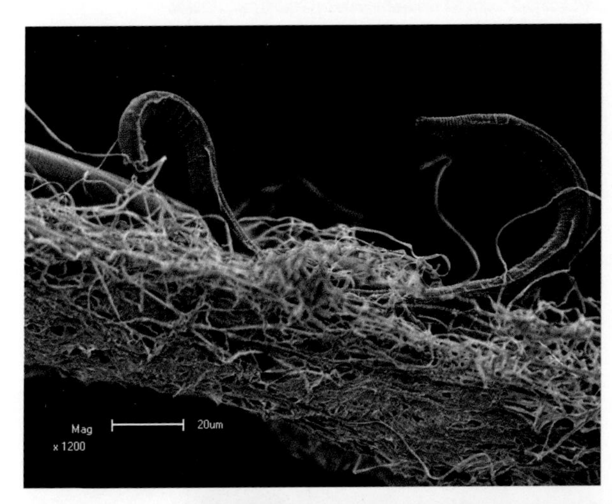

Tiny roundworms (nematodes) collected at a depth of 0.86 miles within Kopanang gold mine, South Africa. GAËTAN BORGONIE

Nematodes inside a stalactite in Beatrix Gold Mine, South Africa, about one mile below the surface. GAËTAN BORGONIE

Geobiologist Magdalena Osburn in the Sanford Underground Research Facility.
STEPHEN KENNY, SURF

An underground lake in Lechuguilla Cave, New Mexico, where researchers have found microbes that turn rock into soil and carve limestone chambers.
MAX WISSHAK

Aerial view of Pleistocene Park, an experimental nature reserve in Siberia. Notice how the areas within the reserve, where grazers roam, are grassier than those outside its fenced borders. NIKITA AND SERGEY ZIMOV

Some of the bison that scientists have brought to Pleistocene Park to help recreate and maintain grasslands. NIKITA AND SERGEY ZIMOV

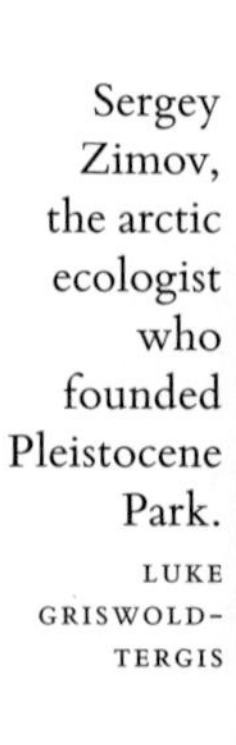

Sergey Zimov, the arctic ecologist who founded Pleistocene Park.
LUKE GRISWOLD-TERGIS

Nikita Zimov, Sergey's son and collaborator, holds a young musk ox.
LUKE GRISWOLD-TERGIS

Network of symbiotic fungi and plant roots spreading through the soil, linking larch and pine seedlings.
ROGER FINLAY

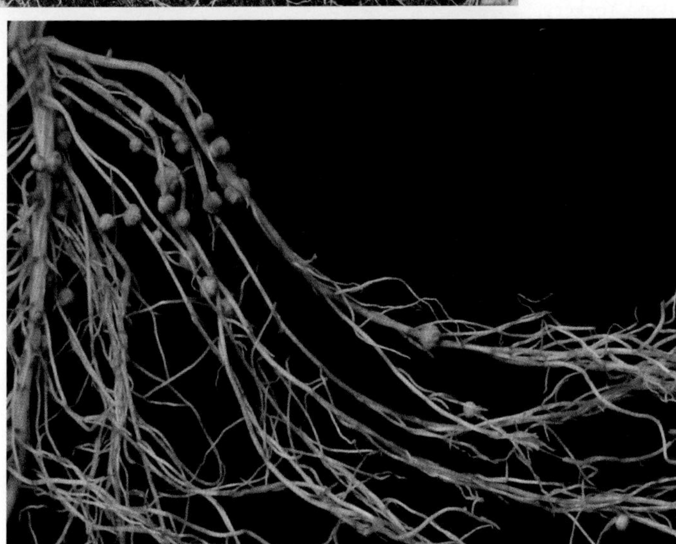

The nodules on these soybean roots contain nitrogen-fixing bacteria, which convert gaseous nitrogen into more biologically useful forms.
BO REN AND JIANXIN MA

The author's backyard in August 2020, when the only vegetation was crispy turfgrass and the soil ecosystem was suffering.
COURTESY OF THE AUTHOR

The author's revitalized garden two years later, with wildlife pond, rockery, drought-tolerant perennials, and raised beds for herbs and vegetables.
COURTESY OF THE AUTHOR

A closer view of the author's garden in midsummer 2022, featuring peak bloom in the rockery.
COURTESY OF THE AUTHOR

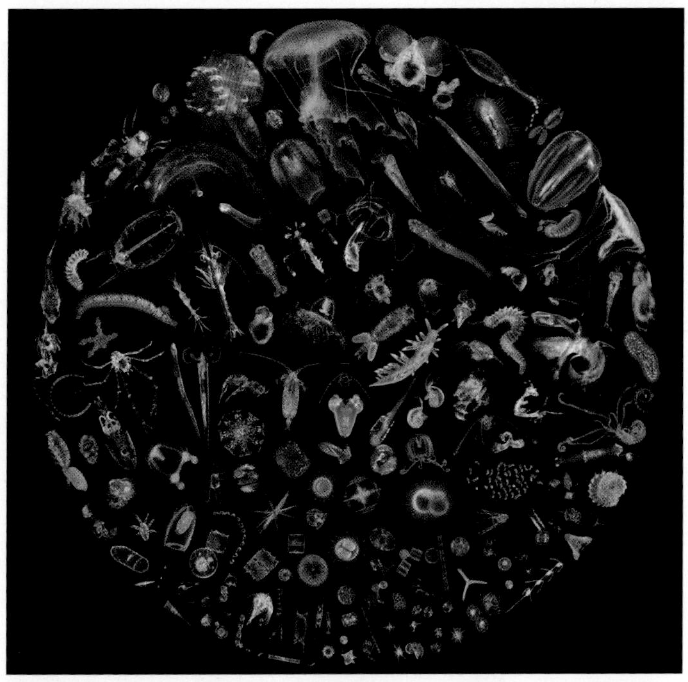

More than two hundred types of the drifting ocean creatures collectively known as plankton, ranging from single-celled diatoms and dino-flagellates through small crustaceans and mollusks to sizeable jellyfish. CHRISTIAN SARDET/PLANKTON CHRONICLES

Single-celled ocean plankton called coccolithophores encase themselves in scaled husks of chalk (calcium carbonate). JEREMY YOUNG

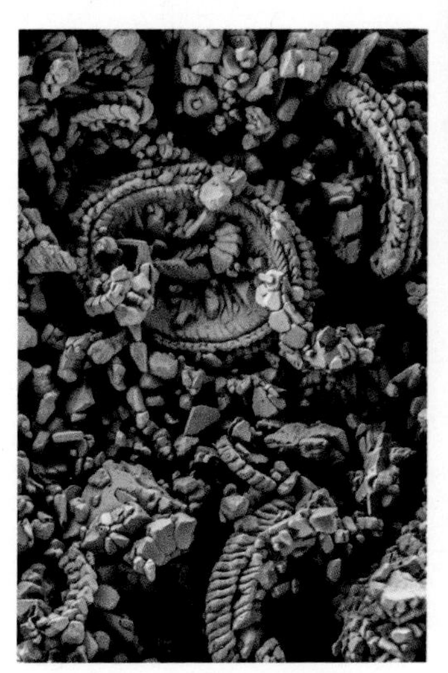

A piece of chalk at high magnification. The White Cliffs of Dover and other chalk formations are made of the fragmented remains of coccolithophores. SCOTT CHIMILESKI

German naturalist and illustrator Ernst Haeckel was mesmerized by the exquisite skeletons of plankton called radiolarians.

BELOW: French architect René Binet's entrance gate for the 1900 world's fair in Paris was inspired by Haeckel's drawings of radiolarians.

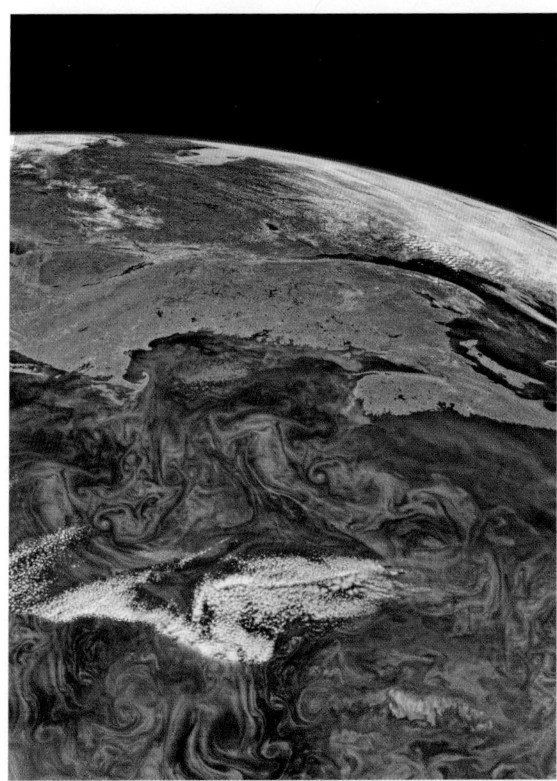

Enhanced satellite image of a massive phytoplankton bloom in the North Atlantic Ocean, along the East Coast of the United States.
LIAN FANG

Kelp forest off the coast of La Jolla, California.
CAMILLE PAGNIELLO, CALIFORNIA SEA GRANT

21st century
Every year, humans emit 60 to 120 times more CO_2 than all volcanoes combined

8,000 years ago
Invention of the plow

Early 20th century
Invention of the Haber-Bosch process

23,000 years ago
Humans experimenting with small-scale grain cultivation

TODAY

4,000 years ago
Humans already burning coal for heat

50,000 years ago–Present
Humans hunt most megafauna to extinction

750,000–300,000 years ago
Homo sapiens emerges

55 mya–Present
Atmospheric oxygen level stabilizes, hovering around 21%

1 million years ago

Cenozoic

66 MYA

240–66 mya
Age of dinosaurs

100–25 mya
Grasses emerge, coevolve with fire, create fertile soils of world's breadbaskets

480–260 mya
Series of ice ages, possibly due in part to land plants

Mesozoic

200–150 mya
Calcifying plankton transform ocean chemistry

PHANEROZOIC

251.902 MYA

394–299 mya
Forests become widespread. Coal deposits form

Paleozoic

420 mya
Earliest fossil evidence of wildfire

541 MYA

540–520 mya
Cambrian Explosion/ Substrate Revolution: Rapid diversification of animals. Burrowing animals irrigate seafloor

500–400 mya
Land plants increase atmospheric oxygen and thicken ozone layer

THE EVOLUTION OF OUR LIVING PLANET

.54 billion years ago (bya)
arth forms

4.4 bya
Molten surface cools. First clouds and rain

4.3–4 bya
Early ocean

4.2–3.5 bya
Possible origin of life

4–3 bya
Continental crust forming, possibly aided by microbes

3.48 bya
Earliest fossil evidence of stromatolites (microbial mats)

3.4–2.5 bya
Cyanobacteria evolve oxygenic photosynthesis

700–425 million years ago (mya)
Fungi, lichen, and eventually plants populate land and accelerate water cycle, rock weathering, and soil production

2.45–2.25 bya
Great Oxygenation Event: Cyanobacteria oxygenate atmosphere, ozone layer forms, sky begins to turn blue

0–580 bya
›ssible series of
›owball Earths

1.6–1 bya
Lineages that become plants, animals, and fungi diverge

2.4–2.1 bya
Global glaciation, possibly due in part to accumulating oxygen

2–1.8 bya
Origin of mitochondria and chloroplasts via endosymbiosis

›strated timeline by Matthew Twombly and Ferris Jabr

A harbor seal swims through giant kelp near Laguna Beach, California. ALEX COWDELL, CALIFORNIA SEA GRANT

Tangled masses of ropes, nets, and fishing gear washed up on Kamilo Beach, Hawaii. SARAH-JEANNE ROYER

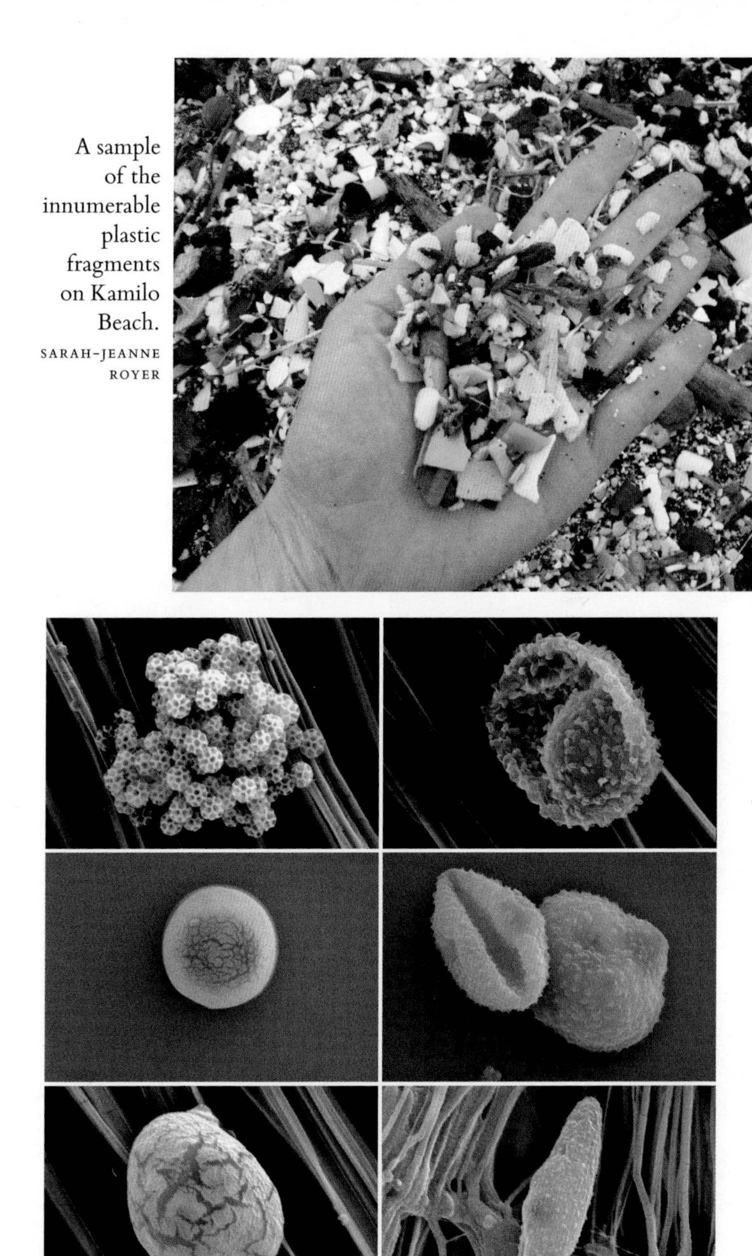

A sample of the innumerable plastic fragments on Kamilo Beach. SARAH-JEANNE ROYER

Tiny airborne biological particles (bioaerosols) collected above the Amazon rainforest, including microbes, fungal spores, and honeycomb-like brochosomes. SACHIN PATADE

Stretching 1,066 feet above the Amazon rainforest, this research tower, part of the Amazon Tall Tower Observatory (ATTO), is the tallest structure in South America.
SEBASTIAN BRILL/MPI-C

Aerosol scientist Christopher Pöhlker at ATTO.
DOM JACK, MAX PLANCK INSTITUTE FOR CHEMISTRY

Atmospheric scientist Cybelli Barbosa looks down on the Amazon's canopy from the tallest tower at ATTO.
RODRIGO ALVES

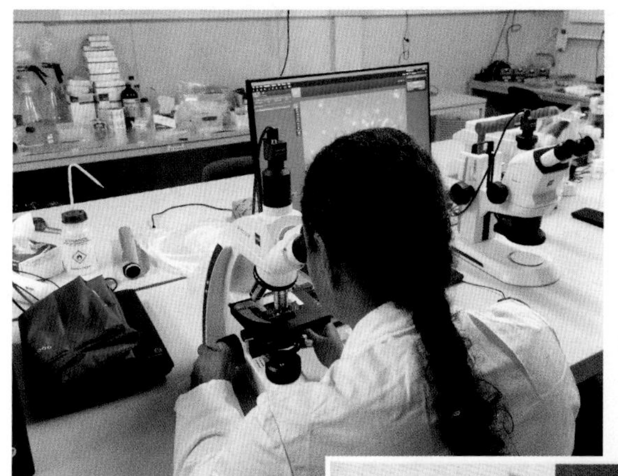

Cybelli Barbosa working in the Clean Lab at ATTO, where biological samples are protected from contamination. SEBASTIAN BRILL/ MPI-C

LEFT: *Pseudomonas syringae,* a type of bacteria that can freeze water. RIGHT: An ice crystal seeded with *P. syringae*. The arrow points to a single bacterium. RUSS SCHNELL

In this Early Devonian scene from 407 million years ago in what is now Scotland, early land plants grow around geyser-fed streams filled with mats of cyanobacteria and algae. VICTOR O. LESHYK

A wildfire rages through a forest in the Carboniferous Period, 300 million years ago. Wildfires were not possible before photosynthetic life oxygenated the atmosphere. WALTER B. MYERS

This geodesic dome on a geothermal plant in Iceland houses a borehole where captured atmospheric carbon is injected deep underground. COURTESY OF THE AUTHOR

Wind turbines spinning beneath a blue sky at the National Renewable Energy Laboratory's Flatirons Campus near Boulder, Colorado. JOSHUA BAUER/NREL

Plastics harm life in many different ways. Ingestion and entanglement are two of the most common. Researchers have documented more than 340 species tangled in abandoned fishing ropes, nets, and other plastic waste, including 26 percent of seabird species, 46 percent of marine mammal species, and every known sea turtle species. Many snared animals drown or suffer gruesome deformities: plastic can become a permanent garrote that slices into the neck of a seal or a girdle that forces a turtle to grow into a figure eight. Scientists have likewise documented ingestion of plastic in more than 2,200 ocean species, ranging from zooplankton to apex predators, including all sea turtle species, nearly 60 percent of both whale species and seabird species, more than one third of seal species, and numerous types of fish. Some animals appear to be attracted to plastic debris in part because it absorbs scents usually associated with their food. Sea turtles and seabirds often consume krill and other crustaceans, which eat plankton and algae, which release pungent dimethyl sulfide, especially when agitated. Birds and turtles have learned to track that odor to find their prey. Seawater-soaked plastic coated with plankton and algae misleads their senses.

The consumption of plastics is further problematic because they often contain and collect toxic substances. Many plastics are insoluble in water and chemically inert and therefore are not particularly toxic, but their molecular building blocks are, posing a risk as plastics degrade. Manufacturers frequently mix plastics with various toxic or otherwise harmful substances to enhance their appearance and performance, including colorants, lubricants, flame retardants, antimicrobials, fillers to reinforce structure and reduce material cost, carbon fibers to increase tensile strength, and plasticizers to improve flexibility and durability. Plastic debris can collect and concentrate environmental pollutants up to one million times higher than concentrations in surrounding seawater. Nanoplastics, which were not even discovered until 2017, are particularly effective sponges of toxic compounds because they have such a high surface-area-to-volume ratio: they are usually less than one micron in size, or about eight times smaller than a red blood cell. Nanoplastics also excel at slipping through the intestines and past the body's defenses, infiltrating blood vessels, the brain, and the immune system.

Minute particles of plastic, and the pollutants they carry, are accumulating in the tissues of people and wildlife alike. Scientists have estimated that, on average, American adults ingest and inhale between 94,000 and 114,000 tiny plastic particles every year, but these figures are considered "drastic underestimates" constrained by limited data. Researchers have even found alarming concentrations of toxic additives from plastic waste in the guts of deep-sea crustaceans living in the Mariana Trench, more than six miles below the surface. Numerous studies have demonstrated that ingested plastic harms the health of zooplankton, mussels, crabs, fish, seabirds, and other organisms by interfering with feeding and reproduction, inhibiting growth, damaging cells, causing inflammation, and modifying gene expression. When researchers exposed perch eggs to concentrations of microplastics typical of coastlines around the Baltic Sea, for example, the larvae that hatched were developmentally delayed, failed to display typical predator avoidance, and died at an unusually high rate.

One of the most troubling ways in which plastics harm ocean life, and the living planet as a whole, is their tendency to disrupt plankton ecology. Mats of floating plastics can prevent sunlight from reaching photosynthetic plankton, inhibiting their metabolism and reproduction. Phytoplankton that cling to plastic debris, or are otherwise exposed to it, can absorb its toxic components. When plastic populated with plankton, algae, and shellfish becomes too heavy and begins to sink, phytoplankton are plunged into darkness. Meanwhile, zooplankton routinely consume toxic microplastics, reducing their growth and fertility.

Plastics also impede plankton's ability to transfer carbon to the deep sea, thus undermining biogeochemical cycles that help regulate the planet's temperature and climate. Most plastic waste floats because the vast majority of modern plastics are less dense than water. When zooplankton eat plastic, their fecal pellets sink more slowly and break up more readily than they would otherwise, reducing the flow of carbon to the depths. Conversely, the continual dusting of microplastics on the seafloor is an eerie echo of marine snow, interjecting a significant and entirely novel source of carbon into seafloor sediments with unclear consequences. "Research into these impacts is still in its infancy, but early indications that plastic pollution may interfere with the largest natural carbon sink on the planet should be cause for immediate attention and serious concern," concluded the Center for International Environmental Law in a 2019 report. Most plastic is made from petroleum, which is itself composed of the remains of plankton and other sea life. Microplastics, then, are a kind of necromancy: long-dead plankton, resurrected, exploited, and eventually discarded in their former home, where they are fated to become ecological impostors, tormenting their living descendants and disrupting the planet's vital rhythms.

IN THE PAST THREE billion years, the Earth system has confronted and assimilated life's problematic waste products many times. Might it do the same now? Will living organisms and the ecosystems they share adapt to humanity's deluge of plastic?

To some extent, they already are. Hundreds if not thousands of species spend at least part of their life cycles floating atop the ocean's surface or drifting just beneath it. Much like plastic debris, such organisms are subject to the currents and are often concentrated in the same parts of the sea. For these creatures, plastics have become both a burden and an opportunity. When scientists used nets to filter seawater throughout the Great Pacific Garbage Patch, they discovered a massive floating ecosystem composed of fish, snails, slugs, crustaceans, and various gelatinous organisms living in some of the highest densities ever recorded. For many of these organisms, plastic is a physical impediment and a toxic pollutant, but for some, plastic can also be a lifeboat or even a home.

The influx of durable floating plastic waste to the ocean is tantamount to the sudden introduction of an immense new habitat where typically coastal species can form self-sustaining communities on the open sea. Scientists have known for centuries that living creatures sometimes reach new territory by rafting across the ocean on wood, seaweed, pumice, and other flotsam. Plastics have dramatically expanded the range and duration of such voyages. The 2011 Tōhoku earthquake and tsunami produced the largest rafting event on record. Following the disaster, researchers discovered that hundreds of species from coastal Japan—including anemones, sponges, and crustaceans—had traveled more than 3,700 miles across the Pacific, primarily by riding on plastic debris. Many of these plastic-bound creatures survived and reproduced in the open ocean for years, eventually arriving on the shores of the Hawaiian Islands and the western coast of North America.

Microorganisms living on mats of floating plastic may also find themselves at an advantage compared to their free-swimming peers because they can more easily feed on each other and one another's byproducts. The biological films that develop on plastic trap potentially nutritious particles, and plastics themselves can be a source of sustenance for microbes that manage to break their tenacious molecular bonds. Some evidence suggests that, over the past seven decades, numerous species have been evolving to do just that. As early as the 1970s, scientists discovered fungi that could break down polyester and bacteria that could digest some of the molecular building blocks of nylon. As of 2020, researchers have documented more than 430 species that can digest various forms of plastic. Most are bacteria or fungi, but there are also some insect larvae in this growing group of plastivores.

In the mid-2010s, a team of scientists in Japan led by microbiologist Kohei Oda collected 250 samples of sediment, soil, wastewater, and activated sludge from a plastic bottle recycling site in Osaka. All of the environmental samples were from areas that had been thoroughly contaminated with polyethylene terephthalate (PET), the main plastic used to make drink bottles. From one of the sediment samples, the researchers isolated a previously unknown species of bacterium capable of digesting PET with two different enzymes and using PET's molecular subunits as its primary source of energy. They named it *Ideonella sakaiensis*. Since then, other researchers have improved the enzymes' efficiency by tweaking their structures and yoking them together "like two Pac-men joined by a piece of string," as John McGeehan, a structural biologist at the University of Portsmouth, once phrased it.

These types of discoveries and advances have fomented interest in the possibility of using plastic-eating microbes to revolutionize the plastics industry. The French biochemical company Carbios is one of several startups trying to create a new, closed-loop recycling system for plastic. Conventional mechanical recycling generally involves crushing old plastic into flakes, melting it down, and remolding it into new products. The process, which typically produces lower-quality plastic, can be repeated only so many times before the plastic must be landfilled. In contrast, Carbios and similar companies hope to use microbial enzymes to reduce plastic to its molecular building blocks and reassemble them into high-quality virgin plastic in an infinite loop. Researchers at Carbios, in collaboration with scientists in Europe, have engineered an enzyme that they claim can break down the equivalent of a hundred thousand flaked plastic bottles within ten hours.

Carbios has already partnered with several major corporations, including PepsiCo, L'Oréal, and Nestlé, and plans to open a 44,000-ton-capacity commercial facility sometime in the next few years, but enzymatic recycling must overcome several hurdles before it becomes a viable industry. Although recycling plastic with enzymes may require less energy and produce fewer greenhouse gas emissions than the production of virgin plastic from fossil fuels, the latter is still about half as costly. PET may be amenable to breakdown by microbial enzymes, but Styrofoam, PVC, and other types of plastic with even stronger molecular bonds may not be. And enzymes are often fussy, requiring highly specific temperatures and pH levels for optimal performance. If enzymatic recycling becomes more efficient and less expensive, it may one day be an important part of a portfolio of tools used to manage plastic waste. Nevertheless, some experts argue that such ambitions should not distract from what is currently possible. There are already chemical methods of reducing PET to its molecular building blocks and reassembling them, for example. And through a combination of environmental taxes on manufacturers and deposit systems that reward consumers, several countries in Europe and Asia—

including Norway, Sweden, Finland, Germany, and Japan—already recycle between 86 and 97 percent of their plastic bottles.

The evidence suggesting that microbes, fungi, and other organisms are already evolving to digest plastic invites an all-too-tempting line of thought: that our living planet will "solve" the problem of plastic pollution on its own. It will not—at least not in any time frame relevant to human society. Just as we cannot afford to wait anywhere close to the time required for Earth to restabilize its climate by itself, we cannot sit back and expect the planet to clean up our mess. Most of the plastic-eating enzymes in the wild work much more slowly than their engineered counterparts. And when microbes break down plastics in the ocean or on land, they aren't necessarily benefiting their ecosystems in the same way as the more familiar processes of decomposition that evolved over millions of years. Instead, especially in the near future, plastic-munching microorganisms may end up creating more nanoplastics, releasing toxic additives into the environment, and adding CO_2 to the atmosphere. Spraying engineered microbes onto plastic waste to accelerate decomposition might be helpful in precisely controlled scenarios, but it could also prove disastrous, like so many similar experiments before it. In the 1971 science fiction novel *Mutant 59,* the world begins to fall apart—literally. Electrical insulation melts, computer networks crash, spacecraft explode, and airplanes dissolve mid-flight. The trouble seems to be specific to the plastic components of various technologies. At first, people suspect that a widely used polymer is at fault. Eventually the truth is revealed: mutant bacteria have escaped a lab and overrun the planet—bacteria engineered to eat plastic.*

WHEN NATTAPONG NITHI-UTHAI WAS a child, plastic was his plaything. Not just plastic toys and figurines but the kind of raw materials used to make them, too. Nithi-Uthai, who goes by the nickname Arm, grew up in Pattani, Thailand, not far from the beach, where his family ran a latex-processing business. He often tinkered with scraps of rubber and

* The full title of the novel is *Mutant 59: The Plastic-Eaters,* by Kit Pedler and Gerry Davis.

other polymers, curious about their form and function. At the same time, he fell in love with the sea. "I feel very connected to the ocean: the smell, the sound, the weather," Arm says. "In Thailand, there are two groups of people: those who like to go to the beach and those who like to go to the mountains. I never went to the mountains—mosquitoes and all that. I am a beach person."

In the 1990s, after completing his undergraduate education in Bangkok, Arm moved to Cleveland, Ohio, to earn a PhD in macromolecular science from Case Western Reserve University, home of the nation's oldest stand-alone polymer science department. When he returned to Pattani in his thirties, parts of it looked different than he remembered—especially the coast. "I lost connection with the ocean when I went away to study," Arm says. "When I came back, I started to see trash everywhere on the beaches." It seemed that, in his absence, Thailand's throw-away culture had intensified. Pollution was more rampant than ever.

In his new job as a researcher and lecturer in the department of polymer and rubber technology at Prince of Songkla University, Arm became increasingly interested in how to reprocess discarded rubber and plastic into something valuable. Used flip-flops were particularly common on Pattani's beaches, comprising 10 to 15 percent of the waste by weight, Arm estimated. Most of those flip-flops were made from rubber foam and plastic straps that could not be melted and remolded like plastic bottles. With a few students, he experimented with methods of grinding flip-flops into small pieces, pressing and binding them into sheets, and shaping them into new products, such as floor tiles and exercise mats. They had some success, but with so few people, they could collect only small amounts of trash.

One day in 2015, Arm noticed a Facebook post about a nonprofit organization called Trash Hero, which was establishing groups of volunteers to clean beaches throughout Asia on a weekly basis. He immediately recognized an opportunity. Within three months, Trash Hero had provided Arm with one hundred thousand flip-flops, which he stored in his backyard in a waist-high heap more than eighty feet long. At first, Arm and his students struggled to turn the pile of garbage into a viable commercial product. Some of the ragged and mismatched flip-flops remained in Arm's yard for so many months that snakes began to nest among them. Toward the end of the year, however, they tried something different—a more literal rebirth. Instead of trying to make entirely distinct products from old flip-flops, they began stamping soles for new flip-flops from the rubber sheets they produced. They called their project Tlejourn, Thai for "sea journey." Local media publicized their work, which attracted the interest of a department store, prompting hundreds of orders.

Not long after, Arm helped found a new chapter of Trash Hero in his community, which is still active. Volunteer cleanups continue to provide Arm and his colleagues with piles of discarded flip-flops. As of late 2022, Tlejourn has made and sold around fifty thousand recycled shoes of various kinds. They produced thirty thousand in partnership with Nanyang, one of Thailand's biggest footwear companies. The rest were made by a small collective of women in a rural village not far from Pattani whose members are skilled in the production of handmade clothing and accessories. Thanks to this collaboration, some of them now have substantially higher incomes.

"If you work with recycled material for a while, you start to see junk as a resource," Arm says. "Our story is simple: we take old flip-flops and give them a new life. Anyone can take a pair of flip-flops anywhere. Our product and philosophy travel with our customers. For the environment as a whole, maybe this is just a drop in the ocean. But our work is more than cleaning up trash. We help people see the larger problem. We help them find in themselves the power to make a change."

In their current state, the systems we use to manufacture and dispose of plastic are inextricably linked not only to global warming but also to the current planetary crisis at large. Extracting fossil fuels and refining them into plastics are extremely energy-intensive processes. Globally, plastics contribute an estimated 4 percent of greenhouse gas emissions—more than air travel. Of the 8.3 billion tons of plastic generated since the 1950s, between 75 to 80 percent has become waste. Only 9 percent of that waste has been recycled; the great majority either has been dumped in landfills or is polluting the ocean. To preserve a livable planet, we must transform our relationship with plastic.

Managing the plastic crisis depends on four essential tasks: dramatically reducing the use of disposable plastic, expanding and improving recycling systems, preventing plastic waste from reaching the ocean in the first place, and removing as much of the plastic that is already there as is feasible. Of the four, the latter has received a disproportionate amount of attention, yet many scientists and environmentalists argue that it is the most fraught and least effective strategy. "Turn off the tap" is a favorite mantra among experts on marine pollution. If your house were flooding due to an overflowing bathtub, they ask, what would you do first: grab a mop or turn off the tap? Clearly, it makes more sense to stop the flood at its source before cleaning up the mess.

Perhaps the most famous attempt at sweeping the seas is Dutch entrepreneur Boyan Slat's nonprofit The Ocean Cleanup, which traps plastic debris in the open ocean by towing a large U-shaped net between two ships. Since the organization's founding in 2013, researchers have criticized it for numerous reasons, including inefficiency, impracticality, outsized carbon emissions, and endangerment of floating sea creatures. The Ocean Cleanup has repeatedly said that it aims to "remove 90% of floating ocean plastic by 2040." As of 2022, the organization claims that it has removed just over 110 tons of plastic from the Great Pacific Garbage Patch—only 0.1 percent of the total in that part of the ocean alone. As marine biologist Rebecca R. Helm has pointed out, certain groups of less renown, such as the Ocean Voyages Institute, have removed several times as much plastic with far less funding, much simpler equipment, a smaller carbon footprint, and less risk to wildlife.

Even the most commendable cleanup efforts on the open sea do nothing to stymie the influx of plastic pollution. Intercepting plastic in rivers and removing it from coastal sites of aggregation, such as beaches and harbors, are increasingly popular strategies. Hundreds of dredgers, skimmers, and workboats designed by the UK-based shipyard Water Witch have collectively retrieved two million metric tons of marine debris from harbors and waterways around the world. A family of semiautonomous, sun-and-water-powered trash conveyor belts that resemble giant roly-polies with googly eyes—known as Mr. Trash Wheel, Professor Trash Wheel, Captain Trash Wheel, and Gwynnda the Good Wheel of the West—continually pull hundreds of tons of refuse from rivers and streams throughout Baltimore, Maryland, each year. As of this writing, The Ocean Cleanup has deployed trash interceptors in nine rivers in Asia and the Americas. In the flooded house analogy, these efforts are like placing a series of sponges and buckets around the overflowing bathtub to catch some of the water before it spills onto the floor—helpful, but still not addressing the source of the problem.

Ultimately, managing the plastic crisis will require a dramatic reduction in the manufacture of disposable plastic and much stricter regulation of unavoidable waste. A 2020 report by the Pew Charitable Trusts concluded that it is possible to reduce the flow of plastic to the ocean by about 80 percent by 2040 through immediate widespread implementation of existing solutions. Taxes and bans on single-use plastics help, as does the elimination of superfluous plastic packaging and the introduction of sustainable alternatives to disposable plastics. Waste collection and disposal systems in low- and middle-income countries must be improved, and manufacturers must be held responsible for the entire life cycle of their products.

Progress is happening, albeit not nearly fast enough. More than one hundred countries and ten U.S. states have banned plastic bags. To meet its target of zero plastic waste by 2030, Canada is in the process of prohibiting the manufacture, sale, and use of a broad range of single-use plastics, including checkout bags, cutlery, ring carriers, and straws, with exceptions for people with disabilities and medical needs. China, India, and the European Union are also gradually attempting large-scale bans on single-use plastics. In 2022, the United Nations Environment Assembly agreed that by 2024 it will begin negotiations on a legally binding international agreement to end plastic pollution and establish a science-policy panel analogous to the Intergovernmental Panel on Climate Change (IPCC).

Shortly thereafter, Norway and Rwanda formed the High Ambition Coalition to End Plastic Pollution, now an alliance of thirty-two countries committed to halting plastic pollution by 2040 through a "comprehensive and circular approach that ensures urgent action and effective interventions along the full lifecycle of plastics." Similarly, the Ellen MacArthur Foundation and the U.N. Environment Programme have united more than a thousand businesses, governments, and other organizations comprising more than 20 percent of the plastic packaging market "behind a common vision of a circular economy for plastic, in which it never becomes waste."

In the 1960s and '70s, around the same time that James Lovelock and Lynn Margulis were developing the Gaia hypothesis, some economists wrote the foundational texts of ecological economics, an interdisciplinary field that studies the human economy as a subsystem of the living planet. In parallel, scholars in various fields formalized a concept with ancient antecedents: the circular economy, which bends the traditional linear economy into a loop, extending the life cycles of materials and products as long as possible through sharing, leasing, reusing, repairing, refurbishing, and recycling.

More recently, economists such as Kate Raworth have integrated modern Earth system science and the concept of planetary boundaries into their economic frameworks. A planetary boundary is a limit on the level of disruption the Earth system can tolerate before becoming dangerously unstable. Violating these thresholds—by, say, depleting the ozone layer or acidifying the oceans—threatens the habitability of the planet for human civilization. In a research paper and subsequent book, Raworth envisioned a "doughnut economics" in which humanity remains within a safe space between an inner ring of essential human rights and needs and an outer ring of ecological ceilings. The linear industrial system of the past two centuries, Raworth wrote in 2018, is fundamentally flawed "because it runs counter to the living world, which thrives by continually recycling life's building blocks." Instead, she continued, "we can study and mimic life's cyclical processes of take and give, death and renewal, in which one creature's waste becomes another's food."

From an ecological perspective, the death or dissolution of any individual entity is not a conclusion but a transition—not a loss but an opportunity. Every creature and object in existence—whether a rock, leaf, whale, or rubber flip-flop; whether formed by geology, evolution, or engineering—has a life cycle, even though we may be too short-lived or shortsighted to see it. The challenge before us is to ensure that all the materials we introduce to our world can be expediently recycled by existing systems or to invent new systems to accommodate them. Before we manufacture yet another plastic bag or bottle, we must take seriously the probability that it will smother a deep-sea coral, splinter into a million charlatan plankton, or contribute to a new layer of the rock record. Before we make a shoe, we must account for every step it will take—not just in the next few years but in its entire indefinite journey through the streams and strata of the planet and all the future ages of the creature we call Earth.

IN 2001, WILDLIFE BIOLOGIST Bill Gilmartin learned that a Hawaiian monk seal had given birth on Kamilo Beach—the first record of the highly endangered species breeding on the Big Island in a very long time. Several years earlier, Gilmartin had co-founded the Hawaiʻi Wildlife Fund, a nonprofit dedicated to the protection of Hawaii's native species. When he heard the report of monk seals on Kamilo, he willed his Subaru Forester down the dirt-and-lava-rock trail leading to the beach and hiked the last mile the car could not navigate. He found the seal and her nursing pup, only a few days old, near the surf. Everything above the high tide line was "a solid mass of nets and plastic," Gilmartin recalls. In some areas, the mounds of refuse rose above his waist.

Over the next few months, Gilmartin frequently returned to Kamilo, sometimes accompanied by University of Hawaii students, and set up camp to keep an eye on the seals and make sure that recreational visitors did not disturb them. In 2003, he got a $10,000 grant from the state of Hawaii to support a major cleanup. In the span of two days, Gilmartin and about seventy Hawaiʻi Wildlife Fund volunteers used a few dump trucks and tractors to haul fifty tons of debris off the shore.

Gilmartin and his colleagues have been organizing routine cleanups of Kamilo ever since.

Early one July morning, my partner Ryan and I drove from Hilo, on the Big Island's eastern shore, to the small community of Nāʻālehu, near its southern tip, where we met several members of the Hawaiʻi Wildlife Fund: Beverly Sylva, a net patrol specialist, electrician, and lifelong beachcomber; Jodie Rosam, an ecologist and fieldwork specialist; and Rosam's five-year-old son, Radan. We clambered aboard a utility task vehicle and began trundling down a dirt road through lowland dry forest toward the sea. Along the way, we passed cave systems long used as shelters by weary travelers, ancient trails formed by generations of Indigenous Hawaiians, and walls of carefully stacked lava rock marking the former boundaries of nineteenth-century cattle ranches. About forty minutes into our journey, as we neared the coast, we began to notice a very different kind of artifact: scraps of plastic fishing nets and ropes, red, yellow, or cyan against the dark earth.

We parked nearby, walked through a gap in the masses of naupaka kahakai, or beach cabbage, and stepped onto Kamilo Beach. At first glance, it was surprisingly ordinary: a stretch of sand accented with lava rock and strewn with driftwood. In contrast to shocking photos and videos of trash heaped upon Kamilo several decades prior, the beach now appeared to be relatively clean. As of early 2023, thanks to the persistent efforts of hundreds of volunteers, the Hawaiʻi Wildlife Fund has removed 320 tons of waste from Kamilo and nearby regions of the coast. Gilmartin says the amount of refuse on the beach has never returned to its turn-of-the-century peak. Still, an estimated fifteen to twenty tons of debris continue to wash ashore every year. Without frequent cleaning, the beach will eventually reaccumulate a thick coating of plastic and other trash.

The quantity of visible pollution one encounters on Kamilo today depends in large part on the currents. Ryan and I visited just after a passing tropical storm. Within a few minutes of exploring, it became apparent that, despite the overall improvement in Kamilo's conditions, there was still considerable debris atop the sand and buried within it. Just about everywhere we looked we found multitudes of small plastic fragments mingled with rock, wood, and shell. When we dug, we uncovered even more. Walking up and down the coast, we saw ropes, nets, buoys, and the funnel-like ends of hagfish traps; plastic bottles that once held glue, ketchup, or shampoo; plastic cutlery, buckets, and gas canisters; a small disembodied wheel, perhaps from a cooler or suitcase; a piece of moldering yellow foam clustered with barnacles; some aqua socks and platform shoes; and a child's blue Velcro sandal embossed with *Toy Story* characters. At one point, Ryan handed me a piece of basalt about the size of a fingerling potato, laced with ribbons of milky turquoise and studded with bits of shell and coral. Plastiglomerate. We found more examples not far away, which variously resembled amorphous lumps of taffy, melted boxes of crayons coated in sand and pebbles, and piñatas that had collapsed in on themselves.

Sylva, dressed in paisley shorts, brown hiking boots, and a wide-brimmed hat, marched up and down the beach, pulling huge pieces of rope from the sand and slinging them over her shoulder. Rosam, practically glowing in a neon green T-shirt, used lava rocks as stepping stones to reach plastic tubs and containers bobbing in the surf. Over the years, Sylva and Rosam have found a wide variety of debris on Kamilo, including car parts, refrigerators and freezers, toilet seats, snow tires, oil drums, fluorescent bulbs, and a message in a champagne bottle tossed into the sea during a fundraiser in Maui.

Members of the Hawai'i Wildlife Fund acknowledge that beach cleanups do not address the source of the plastic crisis—that they don't turn off the tap. Most of the waste collected in coastal cleanups around the world is either landfilled or burned for energy. But these efforts do have other meaningful benefits, such as preventing beached debris from washing back into the sea and reducing risks to wildlife. The act of volunteering also changes how people see the world and live within it. "I know it's not the answer, but it is part of the solution," said Sylva, who grew up in a beach house in Oahu in a mixed Hawaiian and Portuguese family with a deep love for the ocean. "When we bring people here, they are touched by it, and they'll come back with bags and do mini-cleanups of their own. Sometimes we find five, ten, even fifty-pound feed bags that people have filled with debris and left for us to collect because they know we are coming by. If you leave the beach clean, I think campers tend to honor that and leave it the same way."

Alongside the plastic, Sylva has encountered all manner of wildlife on Kamilo and neighboring shorelines. She has seen owls, seabirds, seals, pods of whales off the coast, and a humongous Hawksbill sea turtle estimated to be more than a hundred years old. Now and then, she has also found animals injured or killed by ships, fishing gear, and marine pollution, including turtles with deep gashes in their shells from propellers and a dead juvenile sperm whale.

Sylva told me a story recounted to her by her friend Nohealani Ka'awa, who works with the Hawai'i Wildlife Fund and The Nature Conservancy. In the fall of 2021, a fisherman found a dead female dolphin beached on the southernmost point of the island of Hawai'i, a few miles south of Kamilo. She had been tangled in fishing lines in such a way that every time she kicked her tail, she pulled herself further underwater, eventually drowning. After the National Oceanic and Atmospheric Administration examined and cremated the dolphin, they notified Ka'awa, who was born and raised in the area surrounding Kamilo and who had become known in her community as a practitioner of Hawaiian cultural traditions. Ka'awa and her family brought the dolphin's remains to the spot where she had died. There, dressed in kīhei—sashlike garments often worn during ceremonies—they blew pū (conch shells) and called to Kanaloa, god of the deep ocean. They waded barefoot into the water and poured the ashes into the sea. A little while later, while preparing to leave, Ka'awa saw a hō'ailona—a spiritual sign—in the form of a face in the clouds. She took a photo, which she later sent to Sylva. The features were unmistakable: a domed head tapering to a well-defined beak, mouth slightly parted. A dolphin looking down from the sky.

AIR

CHAPTER

7

A BUBBLE OF BREATH

I KNEW THE TOWER WOULD BE INTIMIDATINGLY TALL. I'D SEEN DIZzying photos of its needlelike profile viewed from afar and sweeping footage from drones that traced its full height. Yet it was not until I arrived at the tower's base, stared up at its spare metal skeleton, and confronted the imminent prospect of climbing to the top that I began to wonder if I had made a mistake.

Situated deep within a relatively pristine region of the Amazon rainforest in northern Brazil, the tower is part of a research facility aptly named the Amazon Tall Tower Observatory. Scientific instruments attached to the structure and its shorter siblings continuously collect airborne particles and gases at different altitudes. Researchers from around the world visit the station to study how the rainforest influences both local ecology and global climate.

The largest tower, the one I was standing beneath, is the tallest structure in South America, stretching 1,066 feet into the sky, about the same height as the Eiffel Tower. Its rectangular steel frame is painted in alternating blocks of orange and white, like a giant safety cone. As I imagined the journey to the top, however, *safe* was not the word that leapt to mind.

To ascend the tower, I would have to climb nearly 1,500 narrow steps with large gaps between them while harnessed to a railing by a short cord. The harness made it unlikely that anyone would plummet to their death, but, I was told, someone could break a limb if they misstepped and fell through one of the many openings in the tower's latticelike framework. I'd heard that, on a few occasions, people climbing the tower for the first time had been unable to continue due to overwhelming fear, a situation that can entail convoluted procedures to return them safely to the ground. Several of the scientists I was accompanying repeatedly asked me whether I was afraid of heights. No, I told them. I'd explored cliffs, mountains, and the decks of skyscrapers without difficulty. Okay, they said. But are you sure?

In truth, I was not entirely sure. The tower seemed incomplete: not so much a finished structure ready for human use as the mere suggestion of a staircase in bare-bones scaffolding incapable of masking vertiginous views. I worried that my brain would not fully register the absurdity and danger of the situation until partway up, at which point I would panic and cling to the railing like a cat whose curiosity had outpaced his courage. But I had not traveled all that way to give up without trying. Once I made it to the top of the tower, I would be halfway between the trees and clouds. I had come to this place to learn about the relationship between them. I had come to see the Amazon make its own rain.

On an obligingly sunny morning in April toward the end of the Amazon's rainy season, I joined a group of visiting scientists in a gear-filled container near the base of the tower. We pulled on climbing harnesses furnished with ropes and carabiners and strapped on mandarin orange hardhats. Each of us had a primary lifeline: a thick cord tethering our harness to a four-wheeled trolley designed to roll along the grooves of a curving rail that flanked the tower's staircase. Sipko Bulthuis, a Manaus-based technician with a shaggy flop of blond hair, was the first to start climbing, followed shortly by Uwe Kuhn, an atmospheric chemist from the Max Planck Institute for Chemistry with foggy blue eyes and a tapered soul patch. I was next. I slid my trolley into the spiraling rail and took the first few steps, gripping the banister tightly with one hand and helping the trolley along with the other. Kuhn's colleague Christopher Pöhlker—a tall, soft-spoken, moon-faced man—followed a little while later.

To my surprise, the ascent was immediately and continuously exhilarating. Thrill and wonder easily overpowered fear. Within ten minutes, we reached the rainforest canopy, about 80 to 115 feet above the ground. Here, the distinct features of individual trees—the yellow-flowered crowns of guayacans, the nearly horizontal branches of kapoks—were still discernible. The death metal screams of howler monkeys rumbled through the air, accompanied by the squawks and chitters of macaws. Above us, the sky was bright and blue, save for a few smudges of white in the distance.

By the time we'd climbed half the tower—533 feet—the view was quite different. We could no longer perceive the majesty of any particular tree. Instead, we saw a vast knurled mat of gray and green stretching to the horizon in every direction. From this height, I understood, more clearly than ever before, that each tree was part of an immense living network covering the planet's surface. As we climbed, the trees pulled an invisible ocean from the soil through their roots into their trunks and tissues. The sun drew what the trees did not use through their leaves into the air, where the water vapor condensed on a floating potpourri of dust, microbes, and organic detritus derived from the forest, eventually forming visible clouds. Lakes of shadow now drifted across the canopy, echoing their woolly counterparts, as if to remind us that the trees were in the clouds and the clouds were in the trees—that forest and sky were twinned phrases in the same ancient chorus.

About an hour after we began climbing, we reached the last flight of stairs. Roosting swallows had covered this part of the tower with droppings, which were now as dry and flaky as ash.* Before we could take the final few steps, we had to remove our trolleys from the staircase railing and slide them into a separate rail on the observation deck while remaining attached to the tower via a secondary carabiner. "Always be connected," Bulthuis said as he explained the procedure. *"Sempre tem que estar conectado."*

I made the necessary adjustments and stepped onto the tower's top-

* Wingless wildlife visits the tower, too. Scientists have found snakes in its midst that presumably slithered up the guy-wires.

most platform, which was surrounded by nothing more than a series of thin, X-shaped panels with gaps wide enough to accommodate the average adult torso. I had anticipated that this would be the scariest part of the climb, yet even here I felt surprisingly secure. The heat of the sun was more intense, and the wind fierce at times, but the tower remained steadfast.

Gazing at the unbroken expanse of rainforest beneath us was a decidedly different experience than exploring it by foot. On the ground, I had been overwhelmed by the beauty and lushness of life all around me and the opportunity to examine it in detail: by the ferns and bromeliads frothing on every branch and the intricate tapestries of moss and lichen; by the tantalizing shimmer of a blue Morpho butterfly and the delicacy of a ghost plant's milky flower trembling on a wiry stalk. At more than a thousand feet in the air, the concept of individual organisms began to blur and dissolve. From this vantage point, the forest seemed less like a place or ecosystem in and of itself than the skin, the fleece, of a much larger entity—one whose true scale I was only beginning to glimpse. It felt as though I'd been trapped in a drop of pond water on a microscope slide, confusing a strand of algae for a jungle, and was only now switching places with the eye behind the lens.

We are so used to thinking of the environment governing life's evolution and sculpting its endless forms. Conventional wisdom maintains that rainforests and other highly biodiverse regions of the planet are the result of fortuitous circumstances. Yet just about everything I could see from the top of the tower, I was coming to understand, was to some extent created by life. Most of the tens of thousands of documented animal species in the Amazon, and all those as yet undiscovered, would not exist if plants and fungi had not populated and transformed the planet's land surfaces half a billion years ago. Complex life may never have evolved, let alone emerged from the sea, if single-celled microbes had not started reforming the ocean and atmosphere several billion years earlier. The soil from which the trees below me grew, the rain-heavy clouds preparing to burst, the color of the sky, the air itself—we owe it all to life.

—

THE DEVELOPMENT OF A stable atmosphere was one of the most important events in Earth's infancy. Without sufficient atmospheric pressure, any liquid water on a planet's surface will eventually boil off into space. If young Earth had not retained liquid water on its surface, life as we know it would not exist. Yet it is also true that without life, Earth's water would not be as, well, fluid. A defining feature of our planet today is not simply the presence of water but the simultaneous existence of water in all its possible states—vapor, liquid, ice—and its continuous movement between air, sea, and land. Over time, life became intertwined with the physics that makes this flux possible.

The hidden threads linking life and the atmosphere have fascinated Russ Schnell since his youth. Growing up in rural Alberta, Canada, he witnessed lightning, hail, and torrential downpours every summer. He liked to watch storm clouds form: these great swirling masses of vapor, like whirlpools in the sky, sucking up air, dust, and whatever was too light to escape their pull. As the clouds inhaled, they gradually grew larger and darker, fuming at their continually redefined borders. As a student at the University of Alberta, Schnell—a short, slender man of twenty with thick blond hair and a disciplined mind—spent his summers assisting a group of atmospheric scientists. One of the project leaders tasked him with investigating the formation of hailstones. How exactly did clouds produce such large chunks of ice?

Water that evaporates into the atmosphere will not automatically freeze at 32°F. Pure water can remain liquid to about −40°F. To freeze at higher temperatures, water needs a seed, or what's technically termed an ice nucleus: a tiny particle that acts as a geometric template, aligning water molecules into a highly organized solid crystal. At the time, in 1968, most scientists thought that airborne water vapor condensed on floating particles of dust and soot and that these kerneled beads of water could in turn freeze if the air were cool enough. But no one knew what kind of particles made the best ice nuclei or how embryonic ice crystals grew into hailstones the size of baseballs and grapefruits. Schnell's task was to dissect the very heart of hail and find the mysterious mote that turned cloud water into ice.

Schnell thought back to the hailstorms he had observed as a kid and all the ones he'd witnessed in the intervening years. They always seemed to form over forests and other densely vegetated areas. What if, Schnell wondered, ice nuclei weren't just inert bits of dirt? What if some were spewed out by trees, or vacuumed up from plants by churning storm clouds? When he told the senior scientists about his idea, they smiled as though amused by a schoolchild's naïveté. Trees helped return water to the atmosphere, of course, but apart from that, what could they possibly have to do with clouds, much less hail? Still, if that was the hypothesis he wanted to pursue, he was free to do so.

For weeks, Schnell roamed nearby forests and fields, grabbing handfuls of grass and plucking leaves from poplars, aspen, and conifers. In the lab, he cut off a small piece of vegetation and sloshed it around in a vial of water to capture whatever invisible particles were on its surface. Using a syringe, he pulled some water from the vial and carefully placed dozens of drops on a temperature-controlled copper plate. He then covered the plate with a glass dome and gradually lowered its temperature. If the drops froze before the plate reached 5°F, then he'd know they contained some kind of ice nucleus that was facilitating the formation of ice crystals. They never did.

One summer evening in 1970, in a rush to get to a party, Schnell left a plastic bag containing grass and water on a shelf in the lab and forgot about it. Ten days later, he discovered that the bag was full of a milky white emulsion. The grass had begun to decay. Rather than throwing it out, Schnell decided to test the rotten grass water on the copper plate. To his astonishment, the water froze at 29.66°F—a much higher temperature than anyone had ever reported in similar conditions. Something in that putrid brew—something biological—was turning water into ice.

Schnell moved to the University of Wyoming for graduate school, where he continued his studies on plant-derived ice nuclei. Suspecting that a plant-loving fungus was involved, Schnell asked a colleague in the botany department, Richard Fresh, to take a look at his leaf samples. Fresh discovered that the ice-forming molecules were in fact proteins clinging to the shell of a rod-shaped bacterium called *Pseudomonas syringae* that tended to live in soil and on plants. The proteins mimicked the shape of ice crystals, providing a perfect template to organize free-floating water molecules into a cohesive solid.

On the ground, the bacteria gave plants frostbite, rupturing their tissues in order to access their nutrients. When storm clouds sucked up air and dust from the ground below, they would inevitably pull in various microorganisms as well. Once in the clouds, *P. syringae* and its proteins could seed ice crystals and hailstones. No scientist had ever seriously proposed that a microbe could freeze water, let alone change the weather, yet here was the proof: a bacterial protein that acted like Kurt Vonnegut's fictional ice-nine.

Enthralled by these discoveries, Schnell embarked on a globe-spanning research expedition. First, he traveled west from Canada across the central United States. Then he flew to England and traveled east across Europe and through Russia on the Trans-Siberian Railway. Before returning home, he toured Japan, Thailand, India, Nepal, Iran, and parts of Africa. Skinny and disheveled, Schnell lived frugally, eating and lodging for as little as $100 a month. Whenever he had the opportunity—on the side of a road, in a field, in a grove of trees—he would stop, collect some leaf litter, and store it in a plastic sandwich bag. Back in Wyoming, he tested dozens of samples from all manner of ecosystems and climates. In every single one, he found active ice nuclei produced by *P. syringae* and other microbes.

Ice-making microorganisms, Schnell realized, were not just important for hail formation. Once they got into the atmosphere, they would also increase the chance of rain. Only a small percentage of all clouds grow heavy enough to rain. The vast majority just evanesce. But the presence of ice nuclei can dramatically change the odds. An ice nucleus can start a chain reaction that quickly freezes lots of cloud water, drawing in even more water and swelling a cloud until it bursts. The proteins that *P. syringae* makes are the most effective ice nuclei ever discovered. Schnell thinks they are a crucial component of the water cycle in ecosystems across the planet. "Almost all rain that falls on land, even over the Sahara and along the tropics, is first an ice crystal," he says.

The private sector quickly recognized the potential of Schnell's discovery. By the 1980s, a company called Snomax had patented the process of creating artificial snow using sterilized proteins isolated from massive vats of *P. syringae*. Ever since, ski resorts around the world have relied on microbial proteins to blanket their slopes. In contrast, the scientific community largely ignored weather-changing microbes for several decades, regarding them as an intriguing but trivial aspect of meteorology not worthy of serious research. In recent years, however—as climate change has pushed scientists to reexamine the complexities of the atmosphere, and astonishing new discoveries have come to light—attitudes have started to change.

It's now clear that *P. syringae* is far from the only organism that can turn water into ice. Numerous bacteria, algae, lichen, and plankton—on land and out at sea—produce ice-seeding proteins. Strong winds, updrafts, thunderstorms, and dust storms routinely whisk these tiny creatures into the atmosphere, where they form celestial colonies for weeks at a time before returning to the planet's surface in the very precipitation they stimulate. In doing so, these accidental aeronauts may influence the planet in profound ways that until now have been largely neglected. "The whole concept has definitely gained a lot of traction," says David Sands, a professor of plant pathology at Montana State University. "We need to recognize these microbes as a part—maybe even a major part—of meteorological processes."

The possibility that the atmosphere teems with unseen life has intrigued scientists since the advent of microbiology in the seventeenth century. Antonie van Leeuwenhoek, one of the first people to observe microbes through a microscope, surmised the existence of "living creatures in the air, which are so small as to escape our sight." In the 1800s, while aboard the HMS *Beagle,* Charles Darwin collected windswept dust over the Atlantic that was later revealed to be full of microbes. And in the early 1900s, Fred C. Meier, a plant pathologist with the U.S. Department of Agriculture, convinced Charles Lindbergh and Amelia Earhart to furnish their aircraft with metal cylinders designed to capture microorganisms.

Only in the late twentieth century, however, did researchers begin to regard airborne microbes as more than passive travelers. In 1978, searching for the origins of a *P. syringae* outbreak in Montana wheat fields, Sands flew a Cessna through the clouds above the crops, sticking petri dishes through a porthole. *P. syringae* soon grew in those dishes. In the 1980s, building on earlier research by Schnell, Sands formally proposed the theory of bioprecipitation: the idea that some bacteria disperse themselves through an elaborate rain dance. "At the time, a lot of people thought it was crazy," says Cindy Morris, who works for France's National Research Institute for Agriculture, Food and Environment, and has long collaborated with Sands. But "no one is telling us we are crazy anymore."

In the mid-2000s, Sands, Morris, and their colleagues collected fresh snow on three continents. Nearly every sample contained ice-nucleating microbes, including plant-dwelling bacteria that had made it all the way to Antarctica. A few years later, scientists in Europe and the United States discovered a variety of microbes in the centers of hailstones. Other researchers analyzed cloud water and measured, on average, tens of thousands of bacteria in each milliliter. Using a century of weather data, Morris and a couple of colleagues have also found statistical support for a bioprecipitation feedback loop: the more intense a rainstorm, the more frequent and intense the storms in the days and weeks ahead, presumably because heavy downpours kick microscopic life up into the air.

Scientists once thought that ice-nucleating proteins evolved primarily as a way for *P. syringae* and its ilk to feed on plants and only secondarily as an opportunistic means of air travel. But *P. syringae* is not always harmful to plants and does not live on them exclusively; it's also found in rivers and lakes. Evolutionary relationships among different ice-making bacteria indicate that ice-nucleating proteins evolved at least 1.75 billion years ago, long before the aquatic ancestors of modern plants began to explore the land. Back then, Morris and Sands propose, these proteins probably helped microbes survive freezing water and major glaciations, perhaps by sequestering damaging ice crystals outside their cells.

Over the eons, ocean waves and powerful winds would have carried microbes into the atmosphere, where they would have encountered DNA-warping ultraviolet light, a dearth of food, and the threat of desiccation. Bacteria with ice-nucleating proteins would have enjoyed a huge advantage over those without: a return ticket to the surface. And microbes that could survive long enough to travel great distances would have expanded their range and possibly found more favorable habitats, as the biologist W. D. Hamilton theorized. The types of bacteria that scientists are finding in precipitation today possess a number of skills that may be adaptations to an ancient familiarity with the high life: pigments that act like sunscreen, for example, and the ability to feed solely on molecules commonly found in cloud water. One study even concluded that certain bacteria can reproduce within clouds.

For most of its history—for somewhere between 2 and 3.5 billion years—Earth was an exclusively microbial planet. For much of that inconceivably long era, there were few if any organisms composed of more than one cell. When more elaborate multicellular organisms emerged and populated the sea and land, they did so within a living matrix of far smaller and more ancient creatures. The rise of plants, fungi, and animals profoundly increased the complexity of Earth's ecosystems not simply by introducing larger and more sophisticated organisms but also by spawning countless new relationships between those organisms and their microbial predecessors.

As living creatures linking land and sky—effectively functioning as sponges and pumps—plants developed a particularly close relationship with the water cycle. At the same time, they became canvasses and conduits for their microbial partners. Wherever the confluence of geology and climate offered an abundance of light, heat, and moisture, plants had an opportunity to thrive. Wherever plants thrived, they did so only by associating with fungi and microbes, including those that seeded clouds and induced rain. The warm, wet regions of the continents grew soft and green with leaf and bud. As plants became stronger and taller, they lofted invisible societies of essentially weightless beings, amplifying their presence in the atmosphere. Together, they pulled water from the soil, pushed it into the air, and called it back again.

Schnell's instincts were right: trees had everything to do with clouds.

THE DAY AFTER CLIMBING the tower, I visited the research station's "Clean Lab": a metal container about the size of an elementary school classroom dedicated to studying biological samples that must be protected from accidental contamination. True to its name, the lab was white, spare, and gleaming, furnished with long workbenches and floor-to-ceiling shelves full of chemicals, storage bins, and scientific instruments. A large fridge stood in the back beside a laminar flow cabinet, a hooded workstation designed to prevent contamination of samples with a stream of filtered air.

Cybelli Barbosa, a cheerful atmospheric scientist in her mid-thirties, agreed to show me around. She adjusted her cherry red wristwatch, pulled on blue nitrile gloves, and lined up a series of plastic tubes on the table in front of us. They were stuffed with what looked like dried rosemary and cloves. "Here we have bryophytes," she said, referring to the formal name for mosses, hornworts, and liverworts—the small, woodless descendants of some of the first land plants. "We're studying how they respond to the environment, but also how they change their environment." She explained that she had collected these specimens from the trunk of a nearby tree to which she had attached sensors at different levels to measure variations in temperature, moisture, and light intensity.

"To get a closer look at what's actually happening, we can use equipment like this," Barbosa said, holding up a portable gray device about the same size, and with roughly the same aesthetics, as a video game console from the mid-1990s. "This is a particle sizer. It has an inlet that sucks in air, which we place very close to an organism, like a mushroom releasing spores. The spores flow within the device, pass through an optical chamber, and are caught on a filter—like this." She showed me a white disc that resembled a paper coaster sprinkled with cinnamon. "Just before the particles are trapped, the device counts them and tells us their size and concentration in the air. The number of fungal spores, pollen grains, and other biological particles in the atmosphere at different times is really important because they influence when it rains."

Barbosa first traveled to the Amazon Tall Tower Observatory, or ATTO for short, in 2012, while completing her master's in environmental engineering. At the time she wanted to compare the chemical composition of air in the city of Manaus with air from a pristine environment largely devoid of human influence. The tall tower had not yet been built, so Barbosa climbed one of the smaller 260-foot-high towers to collect her samples. Even that was a daunting challenge. In addition to having respiratory health problems that could leave her short of breath, she was terrified of heights, and the smaller towers tended to sway unnervingly in the wind. But the professional necessity of climbing daily pushed her to overcome her fear. A few years later, she summited the 1,066-foot-tall tower on her first attempt.

The more time Barbosa spent in the Amazon, the more she started to think about shifting her research priorities. What if instead of simply using the rainforest's clean air as a baseline, she investigated its unique properties? What, exactly, was floating up there above the trees? As Barbosa began her PhD at the Federal University of Paraná, she became increasingly interested in bioaerosols, tiny airborne particles and droplets derived from living organisms. In the past decade, Barbosa, her colleagues, and numerous other scientists visiting ATTO from around the world have collectively formed a much more detailed portrait of how these soaring specks and scraps of life help the Amazon rainforest perform its rain dance.

That trees and other plants suffuse the atmosphere with water has been known for centuries. Plants use only a tiny fraction of the water they absorb from the soil. Most of the water they sop up is lost to the air through pores in leaves and other tissues—a process known as transpiration. Forests and other plant-rich ecosystems transport far more water into the atmosphere than would evaporate from the soil on its own. What was not fully appreciated until relatively recently, however, is how all the invisible gases and minute particles generated by a forest dramatically alter the fate of the water suspended above it.

Trees and other plants continually release a variety of gaseous and often pungent chemical compounds into the air in order to communicate within and across species, among other reasons. Think of the intoxicating perfumes of honeysuckle, jasmine, and lilac, which attract pollinators; the sharp scent of pine needles, which likely deters herbivores and pathogens; and the vegetal musk of freshly mown grass, which is thought to be a distress signal. When some of these volatile compounds drift into the atmosphere, they react with oxygen and sunlight and form new, stickier molecules that clump together, providing adequately large surfaces on which water vapor can condense. Plants and fungi also emit potassium-rich salts that indirectly stimulate cloud formation in a similar way.

At the same time, rainforests and other highly vegetated areas emit a mélange of larger biological aerosols: a complex mixture of organisms and organic entities—whole and fragmented; alive, dead, or somewhere in between—including viruses, microbes, algae, and pollen grains; the spores of fungi, mosses, and ferns; bits of leaves and bark; flecks of fur and feather; and slivers of scales of insect shells.* This levitating assemblage of life and its vestiges can seed both clouds and ice crystals, significantly increasing the likelihood of rain and the pace of the water cycle. Above the Amazon rainforest, bioaerosols comprise more than 80 percent of all airborne particles, much more abundant and important for precipitation than dust and soot.

Together, a rainforest and its local atmosphere create a powerful feedback loop: the more it rains, the more the forest grows; the more the forest grows, the more water, cloud-seeding particles, and ice nuclei it lofts into the air; the more quickly clouds form and swell, the

* Among the most beautiful and mysterious of biological aerosols are tiny honeycombed spheres called brochosomes, which insects known as leafhoppers secrete and rub over their bodies, possibly to give their exoskeletons a water-repellent coating.

more frequently it rains. The 20 billion tons of water that gush from the forest to the atmosphere every day exceed even the volume discharged by the Amazon River itself. Based on these findings, scientists have described the Amazon rainforest as a "biogeochemical reactor" that sustains and stabilizes itself, generating around half of the rain that falls on its canopy.

The atmospheric river produced by the Amazon rainforest, teeming with microbes, spores, and biological exudates, does not remain in place. Some of it travels with air currents to distant cities, farms, and ecosystems, especially in the southern part of the continent, including regions that might otherwise dry out. Through complex chain reactions in the atmosphere, the Amazon supplies precipitation to regions of North America, too, such as the Midwest, the Pacific Northwest, and Canada. Other forests around the world provide similar long-range benefits.

Since 1970, humans have destroyed at least 18 percent of the Amazon rainforest—an area larger than France—primarily to clear space for cattle farms. Small-scale deforestation can temporarily increase rainfall immediately over isolated patches of denuded land due to interactions between soil, heat, and moisture in the air, but chronic, large-scale destruction delays the onset of the wet season, shortens its duration, and dramatically reduces overall precipitation. Scientists have determined that deforestation of the Amazon has likely exacerbated some of South America's worst droughts, including water shortages in São Paulo. One study calculated that, were the Amazon to be obliterated, snowpack in the Sierra Nevada could decrease by 50 percent, with disastrous consequences for agriculture in California's Central Valley and thus for the U.S. food supply.

In the final years before his death in December 2021, esteemed American ecologist Thomas Lovejoy, in collaboration with Brazilian Earth system scientist and Nobel laureate Carlos Nobre, wrote a pair of essays warning that the Amazon rainforest was rapidly approaching a catastrophic and potentially irreversible tipping point. They predicted that if humans destroyed 20 to 25 percent of the Amazon, it would lose its power to summon rain. In combination with climate change and human-caused fires, the ensuing drought would transform large swaths of lush forest into arid and degraded scrubland, unleashing billions of tons of greenhouse gases, severely impairing the forest's capacity to store carbon, and altering weather patterns around the world in unpredictable ways. "There is no point in discovering the precise tipping point by tipping it," they wrote. Not only must deforestation cease, they argued, but large parts of the Amazon must also be restored in order to "maintain its essential role for South America and in sustaining the health of the planet."

WHEN LIFE EMERGED SEVERAL billion years ago, it changed much more than the weather. Long before the first forests towered over the continents, before any complex creature crawled out of the sea, microbes initiated an aerial transformation that was much more subtle and indirect than cloud seeding, yet far more profound. Bit by bit, life altered the chemical composition of the entire atmosphere. Life created the air we breathe today.

In Earth's early history, the atmosphere was most likely composed of carbon dioxide, nitrogen, water vapor, methane, and trace amounts of ammonia, with almost no free oxygen (O_2). To our eyes, the sky would probably have appeared a murky orange, at least intermittently, due to the haze of hydrocarbons that formed when ultraviolet light reacted with methane, which was itself primarily generated by microbes adapted to the oxygen-free environment. Today, oxygen comprises 21 percent of the atmosphere, and on a clear day, the sky is a depthless dome of blue. The oxygenation of Earth was protracted, patchwork, and pulsed—an extended revolution that took nearly two billion years, driven by multiple overlapping geological and biological processes. Although scientists still debate the precise chronology, mechanisms, and many of the details of this global transformation, they agree that life was integral to its completion.

Earth's oxygen-rich atmosphere is inextricably linked to what is arguably the single most important evolutionary innovation in the history of our living planet: photosynthesis. As Oliver Morton has written, photosynthesis is "light made life"—a process by which "the light of the sun becomes the stuff of the earth." By performing photosynthesis, organisms capture the energy in light and store it in convenient chemical packages. In contrast to the solar-powered alchemy that takes place in leafy plants familiar to us today, the earliest forms of photosynthesis probably did not require water or produce oxygen. Some scientists have proposed that the first photosynthesizers were deep ocean microbes living near scalding hydrothermal vents 3.4 billion years ago. In lieu of sunlight, they would have relied on the dim glow of magma and superheated water, breathing hydrogen sulfide and excreting sulfur. At some point between 3.4 and 2.5 billion years ago, blue-green microbes known as cyanobacteria developed a radical new version of photosynthesis, which took advantage of highly abundant resources, spinning sunlight, water, and carbon dioxide into sugar and releasing oxygen as a byproduct.

Oxygen atoms are highly reactive, eagerly forming bonds with other elements. Some of the oxygen emitted by early cyanobacteria reacted with volcanic gases as well as with the iron present in seawater and rocks, forming new compounds and minerals. But a portion of their exhalations began to accumulate as free oxygen in temporary and highly localized oases in the primordial sea and air.

Despite their ingenuity, cyanobacteria did not flourish immediately. At first, they were vastly outnumbered by other types of microbes and were likely restricted to shallow, sunlit, nutrient-rich niches where they managed to outcompete their predecessors. Over time, however, they prospered. About 2.4 billion years ago, close to the midpoint of Earth history, oxygen in the ocean surface and atmosphere as a whole began to rise to an appreciable level, though still far lower than today. Around this time, the once-hazy orange sky probably started moving toward the blue part of the visible spectrum. Although this planetary milestone is known as the Great Oxygenation Event, or the Great Oxidation Event, it was not so much a discrete incident as a lengthy transition lasting over 200 million years. Numerous geological and biological processes were likely responsible for this permanent shift in the Earth system, including changes in the composition of volcanic emissions and chemical reactions that allowed atmospheric hydrogen to escape to space, leaving behind an excess of O2. Whatever the exact mix of mechanisms, cyanobacteria were undoubtedly a critical source of accumulating oxygen. One possibility is that tectonic activity altered the cycling and distribution of phosphorus and other nutrients essential for cyanobacteria, granting them a decisive advantage over their competitors, multiplying their populations, and dramatically increasing overall oxygen production.

As these sun-spinning microbes thrived, they inadvertently pushed the planet to unprecedented extremes. Some scientists have surmised that the early puffs of oxygen generated by cyanobacteria precipitated one of the worst mass extinctions in Earth history, though there is no direct evidence. At the time, most microbial life presumably eschewed oxygen because the highly reactive molecule could initiate damaging chemical reactions, destroying cell structures. The newly oxygenated environment may have proved lethal to the vast majority of Proterozoic microbes, which lacked the suite of antioxidant defenses found in modern organisms.

In parallel, cyanobacteria possibly provoked one of Earth's earliest and most severe climate crises—an unimaginably bleak, planetwide freeze. In Earth's youth, abundant greenhouse gases kept the world warm by trapping heat in the lower atmosphere. In three stages between 2.4 and 2.1 billion years ago, however, the planet's temperature plummeted and massive ice sheets spread across the planet, potentially encasing the land and sea from pole to pole, save for a thin belt near the equator—a hypothesized scenario sometimes known as Snowball Earth. The reasons for this crisis are not definitively known, but the oxygen generated by cyanobacteria may have reacted with methane, converting it into carbon dioxide, which trapped far less heat. At the same time, cyanobacteria removed large amounts of carbon dioxide from the atmosphere through photosynthesis. Without its gaseous blanket, Earth froze. Whatever life survived likely huddled in the few warm havens near volcanoes, hot springs, and seafloor vents. During Earth's first possible snowball event, its thermostat likely kicked in, the planet eventually warmed, cyanobacteria rebounded, and oxygen continued to flush the atmosphere. The whole cycle may have reoccurred several times between 750 and 580 million years ago.

For about a billion years after the Great Oxygenation Event, the level of oxygen in Earth's atmosphere and ocean remained a fraction of what it is today. Yet even this whisper of oxygen may have been sufficient to kindle several evolutionary breakthroughs. Although free oxygen was likely lethal to many of the planet's earliest life forms, its buildup was also an unprecedented opportunity for growth and change. Free oxygen, consisting of two oxygen atoms bound together, not only readily forms bonds with other molecules; it also releases a large amount of energy when its own bonds are broken and rearranged. By adapting to breathe oxygen, organisms increased their metabolic efficiency by a factor of nearly eighteen, which may have encouraged the development of more complex, energy-hungry cells, big bodies, and all manner of corporeal embellishments.

During the initial stages of Earth's oxygenation, perhaps between 2 and 1.8 billion years ago, a series of unique biological mergers forever altered the evolution of life on our planet. First, according to a leading theory, a sizable ocean microbe engulfed a smaller, oxygen-breathing bacterium. For some unknown reason, instead of being digested per usual, the bacterium stayed on as a tenant. Each microbe benefited from the relationship, the smaller one receiving shelter and nourishment and the larger one coopting the ability to breathe oxygen when circumstances suited. As the microbes sustained their symbiosis across generations, the smaller one sacrificed more and more of its autonomy, eventually becoming a permanent cellular structure: the first mitochondrion, the so-called "powerhouse" of the cell. Bean-shaped, energy-generating mitochondria are found in the cells of all complex multicellular creatures today.

At some point after cells with mitochondria arose, the theory goes, some of them assimilated cyanobacteria, which gradually became chloroplasts, the tiny green organelles that perform photosynthesis in the cells of plants and algae. All animals, plants, and fungi are ultimately descended from these ancient fusions of microscopic beings. Lynn Margulis was one of the first scientists to formally develop the theory that mitochondria and chloroplasts were once independent microbes subsumed early in the evolution of multicellular life. At first, her ideas were highly controversial—even ridiculed—but she was eventually vindicated by overwhelming genetic and microbiological evidence.

Scientists continue to debate whether the rise of oxygen in the ocean and atmosphere was a prerequisite for the emergence of complex life, an accelerant, or a different type of influence altogether, but many agree that the absence of free oxygen likely constrained the evolution of elaborate cells and large bodies and that an increasing or oscillating level of oxygen probably contributed to the eventual effervescence of diverse animal forms, such as the Cambrian explosion between 540 and 500 million years ago. By that point, oxygen may have comprised as much as 10.5 percent of the atmosphere, about half as much as today.

Historically, evolution has been depicted as linear and branching, like a tree, or cross-linked, like a web. Although those metaphors certainly capture many evolutionary processes, others are much more sinuous—even circular. Again and again, life and environment alter each other through feedback loops. Through their behaviors and byproducts, living creatures make lasting changes to their surroundings that partly determine the fate of their descendants and of other species. Microbes can seed clouds. Forests on one continent can make it rain on another. Breath can sway a planet.

IN THE DAYS BEFORE climbing the tall tower at ATTO, I spent so much time anxiously imagining the journey to the top that I failed to anticipate the unique challenges of the return trip. After several hours on the tower, my companions and I were exhausted, dehydrated, and sore. As we began descending, my legs trembled involuntarily with a permeating fatigue familiar from long mountain hikes. I had to move even more carefully than on the way up, bracing myself with every step.

Above us, a thick white blanket was knitting itself across the late morning sky. Ironically, the very phenomenon that many scientists visit the tower to study often interferes with their work. In a heavy downpour, and especially during a thunderstorm, the tower becomes even more treacherous than usual. I'd been warned that if it started to rain while we were climbing, we would probably have to turn back before reaching the top. Having evaded that disappointment, I was now hoping to avoid the dangers of navigating a slick staircase with limited dexterity. Propelled by a sense of urgency and helped along by gravity, I finally reached the tower's base, drenched only in sweat and immensely relieved to feel the breadth of the earth beneath my feet again.

No more than twenty minutes later, while I was recuperating at the camp, the temperature suddenly dipped, a strong wind began to bend the trees, and the sky split open, unleashing fearsome torrents. Though this was not the first deluge during my trip, it was the most intense. The air became a wet blur of chaotic particles—a liquid white noise. Water spilled from rooftops, streaming and pooling on the sandy soil. The sound was physically overwhelming, as though it were trying to swallow the terrestrial plane; as though the rain had remembered that it was once an ocean and still had the power to submerge the world.

I looked up at the clouds, the source of this inundation. We see clouds so often, and in such abundance, that it's easy to forget what marvels they are. A cloud is ethereal, yet astonishingly heavy: a levitating lake, typically weighing more than several blue whales. A cloud is aerial alchemy, at once liquid, vapor, and crystal: an enigmatic yet inevitable outcome of atmospheric physics. As I now knew, however, clouds are also biological, sprinkled with microbes and spores, strewn with the remnants of life, forming in the ancient exhalations of living creatures. A cloud is Earth seeing its own breath.

Gazing skyward, I recalled the alternate view of the forest from more than a thousand feet in the air: mile after mile of pristine rainforest receding to the thin gray line of the horizon. By volume, every mature tree in a forest is mostly dead tissue: a pillar of lifeless wood laced with thin layers of active cells, frocked with leaves, and sheathed in symbiotic microbes. Yet no scientist disputes that a tree is alive. Perhaps a forest, with its intricate tangle of the animate and the inanimate, is not so different. Most people, I'd venture, would not hesitate to describe a forest as alive. It seems untruthful to claim otherwise, especially now that science has attested the fundamental interdependency of life, air, and soil; detailed how forests generate much of the rain that falls on their canopies; and revealed vast underground networks of roots and fungi through which trees and other plants exchange resources and information. The concept of a living planet goes one step further. It's not that Earth is a single living organism in exactly the same way as a bird or bacterium, or even a superorganism akin to an ant colony, but rather that the planet is the largest known living system—the confluence of all other ecosystems—with structures, rhythms, and self-regulating processes that resemble those of its smaller constituent life forms. Life rhymes at every scale.

The dominant scientific paradigms of the past two centuries have regarded the origin of life as something that happened *on* or *in* Earth, as though the planet were simply the setting for a singular phenomenon—the manger that housed a miracle. But the two cannot be separated in this way. Life *is* Earth. Our living Earth is the miracle. Life emerged from, is made of, and returns to Earth. We still carry the ocean in our blood and grow skeletons of rock. The origin of life was Earth discovering itself, organizing itself, learning new ways to change. Ever since, what we call life and what we call the planet have been as one, each continually consuming and renewing the other. Earth is a rock that broiled, gushed, and bloomed: the flowering callus of a half-sealed Vesuvius suspended in a bubble of breath. Earth is a stone that eats starlight and radiates song, whirling through the inscrutable emptiness of space—pulsing, breathing, evolving—and just as capable of dying as we are.

Almost as suddenly as it began, the rain stopped, though a considerable volume of water continued to percolate through the forest. What had been a deafening onslaught of noise became a meditative dribble, curiously similar to the soft crackle of a fading fire. Although this was an ending, there was an atmosphere of anticipation, too: perhaps not a prelude but a bridge. I watched the glistening leaves sway and flutter beneath the falling drops, each bowing and rising in turn. They really did seem to be dancing, moving to a music I was still learning how to hear.

CHAPTER

8

THE ROOTS OF FIRE

PERCHED ON A DENSELY FORESTED HILL CRISSCROSSED WITH NARROW, winding, often unsigned roads, Frank Lake's house in Orleans, California, is not easy to find. On my way there one afternoon in late October, I got lost and inadvertently trespassed on two of his neighbors' properties before I found the right place. When Lake and his wife, Luna, bought their home in 2008, it was essentially a small cabin with a few amenities. They expanded it into a long and handsome red house with a gabled entrance and a wooden porch. A weathered arbor twined with kiwi frames the front yard, which features a pond, a vegetable patch, and a ring of huckleberries and coppiced hazel. Nearby stand several open pole barns that function as workshops and storage units for Lake, a research ecologist for the United States Forest Service. A maze of Douglas firs, maples, and oaks, undergrown with ferns, blackberries, and manzanitas, covers much of the surrounding area.

"This is a feral orchard," Lake said as he showed me around his property, weaving in and out of slender-trunked trees and sprawling shrubs. He was wearing cargo pants, thick black boots, and a camo print beanie. "This is an old place that Karuk managed." Lake, who is of mixed Indigenous, European, and Mexican heritage, is a descendant of the Karuk, a native people of northwest California and one of the largest tribes in the state today. Some of his family members are also part of the Yurok Tribe, which is indigenous to the same region. Lake grew up learning the history and culture of both peoples.

"How do we know that that big acorn tree and that one over there and these other ones were part of an orchard?" he continued. "Well, it's this big flat area on red clay soil that's not too far from the historical village and trails. And there are artifacts here," he said, his pace quickening as his words tumbled out of him. "I've found arrowheads in the chicken coop that turned up in the rain or were dug up by gophers. And more stuff in the yard over there, like a mortar and pestle for processing acorns. I could see that this was an acorn place, and it needed love."

A little ways ahead, we reached a grove of moderately large oak trees. Here, the forest floor was mostly free of vegetation, charred black in places, and littered with acorns. Bright green moss draped itself lavishly across a decaying log and several stumps near the center of the clearing. When Lake first moved here, this area was a suffocating thicket of oak trees, madrones, poison oak, and honeysuckle, much too dense to see through or navigate. Since 2009, Lake, who is a certified firefighter, has used chainsaws, propane torches, and driptorches to strategically thin and burn this particular half-acre. Over the years, the controlled burns, or prescribed burns as they are often called, have removed all the smothering underbrush, reduced the number of trees, and provided the remaining oaks—the largest and oldest individuals—with much more light and space, creating an orchard similar to those Lake's ancestors would have managed.

Fire has also kept pests in check. Every summer, weevils and moths lay eggs on or within acorns, which their larvae proceed to devour. Periodic low-level fires spare the trees but kill a portion of the pests' pupae buried in leaf litter and soil, preventing them from ruining the following year's crop. Like many Indigenous Peoples in the area, Lake's family and friends continue to use acorns to make flour, bread, and soup.

"If the trees have too much duff and litter around them, they're gonna have high infestation rates, and not many animals will want to eat their acorns, including humans," Lake explained. "When you have healthy acorn trees that are cleared out underneath like these, all the wildlife starts to show up. The squirrels, the deer. My neighbor saw a big fisher in here last spring," he added, referring to a tree-climbing, weasellike omnivore. "I've seen acorn woodpeckers, too. If you burn, you get good acorns."

"How do you know which ones are best?" I said, scanning the hundreds of fallen acorns around our feet.

"Look for silvery white ones," Lake said, stooping and picking up an acorn. "Like this, right?" He examined it more closely. "Actually, that one's a little chewed on." He continued to rummage through the leaf litter, his fingers moving too quickly for me to follow. "These are buggy. These are cracked and already stained. Okay, here we go. Brown top bad. White top good." He showed me several large acorns with neat white circles on their rounded ends, much cleaner and brighter than the ones before.

"White top good," I repeated. "And why is that?"

"A stain on top usually represents that there's a bug hole or injury. When it's clear, then the inside is usually good, too." Lake cracked open an acorn and split it in half lengthwise. The flesh was smooth and creamy white with a tinge of yellow, like French vanilla. "It's a good acorn all the way around," he said, turning it this way and that, as if he were inspecting a jewel. He paused for a moment in admiration. "That is a perfect acorn," he said. "There's a sense of pride in that. I have a lot of pride in my family to know that our acorns go to the ceremonies, to the feeding of elders, and to the critters that use this as a place of productivity. This is what traditional management and food security looks like. This is human services for ecosystems. And this is climate adaptation. If someone tosses a cigarette on a hot summer day and a wildfire comes through here, this clearing will be a barrier between the fire and my home, and between my land and adjacent landowners. This is a place of safety."

When Lake was a child, his family taught him that fire could feed, nurture, and heal—that it was sacred. Like Indigenous Peoples throughout the world, Lake's predecessors intentionally used fire to change their environments in beneficial ways. When European colonists arrived in western North America, they often encountered beautiful, parklike mosaics of forest and grassland, so open and spacious that they could easily maneuver horse-drawn carriages through them. Although colonists mistook such landscapes for pristine wilderness, they were actually the result of thousands upon thousands of years of careful stewardship by Native Peoples—stewardship founded on a deep understanding of the ecology of fire. Before European settlement, controlled fires started by Native Peoples, in combination with lightning-ignited wildfires, burned somewhere between 4 and 13 million acres every year in California alone.

The benefits of fire for Indigenous communities were myriad. By reducing the density of trees and removing underbrush, controlled burns improved visibility, eased travel, eliminated hiding places that approaching enemies might use, and promoted the growth of meadows and prairies, which attracted deer, elk, bison, and other game animals. Some Native Peoples, such as the Yahi and Mono, hunted bears and deer by trapping them in rings of fire and shooting them with arrows. The Yuki and Pomo burned dry fields to gather grasshoppers. Fire felled trees too large to chop and illuminated paths when the moon and stars would not suffice. Smoke signals conveyed messages over distances too great for feet or sound. Indigenous Peoples routinely burned willow, hazel, redbud, and many other plants to stimulate the fresh, pliant growth they needed to make boats, tools, cordage, clothing, and baskets. A single cradleboard, for example—a kind of baby-carrier—might require 500 to 675 supple sticks from six different patches of fire-managed sumac. Deliberate burning killed or displaced snakes, rodents, ticks, and fleas, while simultaneously spreading fertilizing layers of ash, accelerating nutrient cycles, and encouraging the production of edible tubers, mushrooms, seeds, and berries. By judiciously burning the land surrounding their settlements, Native Peoples also reduced the risk of severe wildfires and protected themselves from encroaching blazes.

When Europeans invaded North America, they initially used fire for some of the same reasons. Over the centuries, however, colonists radically altered the ecology of fire in the Americas. Whereas Indigenous burning traditions tended to promote biodiversity and an abundance of useful native plants, settlers often ignited fires to clear vast tracts of land for large-scale monoculture. In their attempt to domesticate unfamiliar landscapes and impose environmental uniformity, colonists disrupted ecological rhythms that Native Peoples had helped establish over millennia.

By the mid-1800s, disease, war, and forced displacement had devastated Indigenous American cultures, including traditional burning practices. As the timber industry boomed and farms expanded through mechanization, settlers became increasingly averse to fire, which they saw as a threat to valuable hardwood and established cropland. Starting in the late eighteenth century, if not earlier, laws mandated the suppression of Indigenous burns. "In the early 1900s, we were shot for trying to use fire as a land management tool," says Margo Robbins, a member of the Yurok Tribe and co-founder and executive director of the Cultural Fire Management Council. In a 1918 letter, Forest Service district ranger F. W. Harley deplored "the renegade whites and indians" who set fires "for pure cussedness or in a spirit of dont care a damativeness," claiming that the only solution was to shoot them like coyotes. "In later years," Robbins says, "there was a policy to imprison people that set unauthorized fires and to put out all fires by ten o'clock the next morning. Those were pretty effective deterrents to traditional burning."

Following a series of particularly destructive wildfires in the early twentieth century, the United States Forest Service began to prevent and extinguish wildfires even more aggressively—a policy eventually epitomized in the pointed plea of Smokey Bear, who remains central to the longest-running public service announcement campaign in U.S. history. In the absence of the periodic, low-intensity wildfires to which they'd been accustomed, many forests became congested, brittle tinderboxes.

This legacy of fire suppression has left California and other Western states especially susceptible to the kind of catastrophic wildfires they have endured in the past two decades. Global warming, prolonged drought, extreme heat waves, invasive insects and fungi, poorly maintained power lines, the continued dispersal of people into remote and densely vegetated areas, and strong air currents particular to the West have also conspired to create the conditions for infernos of unprecedented scale and severity. Since 2000, an average of seven million acres have burned annually in the United States, more than double the average area in the 1990s. In 2020, wildfires scorched more than 10.1 million acres, the second-highest annual acreage since 1960, when accurate records began. The vast majority of those acres were in the West, and around 40 percent were in California. Although the number of acres that burned in western North America before European colonization was far higher, that was largely the result of frequent, low-grade burns, much milder than the terrifying runaway blazes of late. Today's wildfires burn bigger and hotter than any on record. Some are so large and fierce that they generate their own weather, forming fire tornadoes and pyrocumulonimbus clouds that shoot lightning and scatter embers, igniting even more fires. As of this writing, the fifteen largest wildfires in California's recorded history all happened between 2003 and 2021. Ten of these fires burned in 2018 or later. Five occurred in 2020 alone.

To paraphrase Leaf Hillman, director of natural resources and environmental policy for the Karuk Tribe, a historically misguided fear of fire has created a situation in which many people now have genuine cause to be afraid. "We have to re-establish a positive relationship with fire," he said in one interview. "Our first response has to be, 'manage it.' Not 'suppress it,' but 'manage it.' "

In Florida and much of the southeast, where wildfires were never suppressed to the same degree as in Western states, massive, raging wildfires have historically been a much less frequent problem, and prescribed burns remain relatively common today. Although ecologists and foresters have long recognized the repercussions of fire suppression in the western United States, meaningful change to fire management policies and practices has been slow. One of the major difficulties is the West's convoluted patchwork of federal, state, Tribal, and private land, each with its own set of regulations. As climate change stretches the fire season, the windows in which controlled burns can be safely conducted shrink. Even with all the necessary precautions, a small percentage escape and burn where they shouldn't. And though the smoke generated by controlled burns is far less than that of megafires, it can diminish air quality and endanger some people's health.

Nevertheless, prescribed burning is one of the most effective, economical, and ecologically sensible ways to reduce the risk of further catastrophic fires. Confronted with the rapidly escalating wildfire crisis, the U.S. Forest Service, the California Department of Forestry and Fire Protection, and other government agencies have begun to recognize the need to revive prescribed fire in western North America.

"I think we can change our trajectory and make a positive difference," says Scott Stephens, one of the country's foremost experts on fire ecology. "At the same time, I am more worried than ever about the ability to do this work at the necessary scale and pace. You look out the window now and you see things happening that are mind-boggling. I am still hopeful, but I think we need to move fast."

FOR THE FIRST FEW billion years of Earth's history, wildfires as we understand them did not exist. Fire requires three ingredients: fuel, oxygen, and heat. In Earth's youth, there were many sources of intense heat and plenty of sparks—lightning, volcanoes, rockfalls, asteroid impacts—but hardly any free oxygen, nor much dry and combustible matter. By 600 million years ago, cyanobacteria and algae had gradually and fitfully raised the amount of oxygen in Earth's atmosphere to somewhere between 10 percent and half of its current level—a monumental change but not quite sufficient for fire. The creation of a more familiar atmosphere required a second revolution orchestrated by life: the greening of a new Earthly domain.

As long as 700 million years ago, or possibly even earlier, algae began oozing onto land from the sea. The earliest algal pioneers may have inhabited ephemeral ponds and lakes that periodically dried up, encouraging adaptations to drought. Durable spores protected their offspring from desiccation. Roots provided structural strength and access to distant sources of water and nutrients. Internal plumbing circulated liquids and sugars through ever-larger bodies. Leaves greatly expanded the surface area available for photosynthesis.

Somewhere between 500 and 425 million years ago, the first land plants evolved. *Cooksonia* was a tiny, swamp-loving plant with spore-bearing structures that resembled the toe pads of a tree frog. *Baragwanathia longifolia*'s undulating three-foot-long branches, densely packed with slender leaves, gave it a hirsute, tarantulalike appearance. And the twenty-three-inch-high *Psilophyton dawsonii,* which boasted a rather sophisticated vascular system for its time, looked like a primordial cousin of dill. Land plants mingled with other Paleozoic pilgrims—with microbes, fungi, and animals—forging new partnerships. Earth's crust, once rigid, barren, and gray, was beginning to loosen, wriggle, and leaf.

Between 400 and 360 million years ago, plants experienced a burst of evolutionary innovation, developing much larger and more complex bodies, broad leaves, and robust roots. Whereas the earliest land plants were diminutive creatures that foreshadowed today's mosses, hornworts, and liverworts, some plants in the late Devonian Period began morphing into the first trees. *Archaeopteris,** which could grow between eighty and one hundred feet tall, looked like an immense, shaggy Christmas tree with bowed, fernlike fronds. *Lepidodendron* had scaly bark; a dense, fractal crown covered with small leaves; and roots more than thirty-six feet long. *Calamites* resembled gargantuan versions of today's horsetails, reaching sixty feet or higher. Ferns, uncannily similar to their modern counterparts, flourished alongside these new botanical giants, sometimes matching them in size. By 380 million years ago, primeval forests covered vast swaths of the planet.

Not long after, the first seed-bearing plants emerged. They proved immensely successful, diversifying into familiar forms like ginkgos and conifers. Sometime between 250 and 150 million years ago, certain plants evolved a remarkable new feature: uniquely shaped leaves that flagged the location of their pollen, which made them even more at-

* Not to be confused with *Archaeopteryx,* the birdlike dinosaur that lived during the Jurassic.

tractive to pollen-eating insects that had been inadvertently helping the plants reproduce. Soon the leafy flags adopted bold colors, which helped them stand out in a tangle of green. Perfume and nectar sweetened the deal. By 65 million years ago, flowers had become a global phenomenon. Around the same time, grasses—which may have emerged more than 100 million years ago—were gradually spreading, eventually covering 30 to 40 percent of the planet's land surface.

Plants profoundly altered Earth's crust and atmosphere. Oxygen from cyanobacteria had already started to form a layer of ozone in the stratosphere, which shielded life from harmful ultraviolet radiation; land plants thickened it, sheltering new waves of terrestrial explorers. The greening of the continents dramatically accelerated the water cycle and thus the rate at which rock was weathered. Plants, fungi, and microbes fractured rocks with roots, dissolved them with acids, and enriched the land with organic matter, turning obdurate crust into supple soil. Trees, shrubs, and other large plants with extensive root systems stabilized riverbanks, encouraged rivers and streams to meander sinuously across landscapes, and prevented them from washing too much of the accumulating soil and mud into the sea.

Belowground, plant roots and fungi formed partnerships known as mycorrhizas: threadlike fungi enveloped and fused with plant roots, helping them extract water and nutrients like phosphorus and nitrogen from the soil in exchange for carbon-rich sugars. As terrestrial ecosystems matured, these symbiotic webs became more complex and robust, allowing trees and other plants to exchange water, food, and chemical messages with one another, too.

Once land plants were established, they helped push atmospheric oxygen to its modern level—and beyond. The process by which this happened was not as simple as plants exhaling oxygen into the air. The vast majority of oxygen that photosynthetic ocean plankton and land plants breathe out is used up by other organisms in a perpetual cycle. In order to grow, plankton and plants absorb carbon dioxide, use it to build their tissues, and release oxygen as a waste product. Animals, fungi, and microbes eat and decompose plankton and plants, using oxygen in the process and exhaling carbon dioxide. Not all photosynthesizers are consumed or decomposed, however. A fraction are buried relatively intact on the seafloor or in lakes, swamps, and landslides. The oxygen that animals and decomposers would have used to break down those absentee plankton and plants remains in the atmosphere, having escaped the usual cycle. Bit by bit, this excess oxygen accumulates.

Over the eons, the sun-powered alchemy of marine plankton and land plants—combined with Earth's ceaseless swallowing of life—raised the level of oxygen in the atmosphere from essentially zero to a peak of around 30 to 35 percent in the Carboniferous Period (358.9 to 298.9 million years ago), followed by a similar peak in the Cretaceous (145 to 66 million years ago). Bathed in dense, oxygen-rich air, which made breathing and flying much easier, Carboniferous insects and arthropods ballooned: millipedes grew as large as surfboards and dragonflies sailed on wings as long as those of modern pigeons.

Land plants also became a critical component of Earth's long-term carbon cycle and thermostat. The collective growth and activity of terrestrial plants, fungi, and microbes degrade rock at least five times faster than rain, wind, and ice alone, drawing carbon from the air in the process and accelerating its burial. By removing a potent greenhouse gas from the atmosphere, this biological weathering, as it's known, tends to cool the planet. During the transition from the Devonian to the Carboniferous, not long after the spread of forests, Earth began to experience another ice age and mass extinction, which might have lasted some 100 million years. Shifting continents and rearranged ocean currents were partly responsible, but trees and other land plants likely played a significant role, too. Around the same time, plants evolved lignin and other tough structural tissues that microbes and fungi could not yet thoroughly decompose. In the words of botanist David Beerling, "global indigestion ensued," and large quantities of carbon were entombed in swamps and peatland, further chilling the planet. Eventually, however, as with earlier global glaciations, Earth's thermostat reset the climate. Over time, symbiotic microbes, fungi, and animals such as termites evolved the ability to digest even the most obstinate plant tissues.

With atmospheric oxygen at a historically high level and vast forests covering the continents, Earth was now more hospitable—more alive—than it had ever been. It was also far more flammable. In this new Earth, fire became a customary phenomenon. The charred remains of a 420-million-year-old plant, preserved in siltstone, are the earliest evidence of wildfire. Charcoal has been present in the fossil record ever since.

From the Devonian onward, many plants gradually adapted to fire's recurring presence. They evolved thick, flame-resistant bark, succulent leaves, and resilient tubers that resurrected themselves in charred soil. Some plants even came to depend on fire to reproduce: certain pine trees have cones sealed by resin that melts in the heat of a wildfire, releasing seeds into fertile ash; smoke seems to stimulate germination in some plant species; and a few flowering plants burst into bloom only after a blaze.

In tandem, fire adapted to life. "Fire cannot exist without the living world," fire historian Stephen Pyne writes in *Fire: A Brief History*. "The chemistry of combustion has progressively embedded itself within a biology of burning." Wherever wildfires became regular occurrences, they began an ongoing process of evolving with the very ecosystems that made their existence possible. The outcome is known as a fire regime: the pattern—the typical frequency, intensity, and duration—of wildfires in a given region. If fire is itself a kind of music that results from the interplay of life and environment, then a fire regime is a tune or theme that recurring wildfires and their particular habitat compose together.

Many of the world's forests have evolved with periodic wildfires of varying intensity. Outside of tropical rainforests, fire and forests tend to regenerate one another. Grasslands, prairie, and savanna have developed particularly intimate relationships with fire: when wildfires create vacancies in certain forests, grasses sometimes move in, adapting to the hot and dry terrain; in turn, all the new vegetation fuels more fires, which deep-rooted and vigorous grasses readily endure and encourage, continuing the cycle. Even wetlands have formed their own alliances with fire.

Once fire became a frequent occurrence in the Earth system, an entirely novel evolutionary path emerged: the chance that one or more creatures might learn to control it. When chimpanzees see a wildfire sweeping toward them across the savanna, they do not always flee. Sometimes they track the fire's progress from a safe distance and, once the flames have passed, cautiously inspect the scorched brush. In other cases, they encounter a site that was burned days or weeks earlier. Among the singed shrubs and ashes, they might find morsels of food: charred seeds and fruits, perhaps, or tender green shoots; abandoned bird eggs; insects and lizards exposed by the denuded landscape or scorched before they could escape. Baboons and vervet monkeys also forage in the wake of wildfires. Millions of years ago, early humans probably did, too.

At some point—possibly between one and two million years ago, though no one knows precisely when—our ancestors began manipulating fire itself. Perhaps they started to imitate certain hawks and falcons, which drop burning branches onto fields to ignite blazes and force prey out of hiding. Maybe they transported burning vegetation to rings of stone or simple hearths, where they roasted tubers and learned to use wood, dry grass, and animal dung as fuel. Archaeological evidence—such as thoroughly burned leaves, twigs, and bones as well as patches of soil and flakes of stone heated to high temperatures—suggests that humans were routinely maintaining fires by about four hundred thousand years ago.

Fire was warmth when there was no sun and light when it was not day. Flames warded off dangerous predators, deterred pests, and prevented people from freezing to death. An evening campfire became a focal point of conversation and storytelling. A torch or oil lamp turned the once dark contours of a cave into a canvas for myth and memory. A combination of hunting and cooking with fire allowed our species to evolve and nourish much bigger, denser, and hungrier brains with nearly three times as many neurons. Fire is arguably the single most important catalyst of human evolution—the furnace behind our intelligence, technology, and culture.

Strategically burning the environment is undoubtedly an ancient practice, but its exact origins are lost to unrecorded history. What is certain, however, is that whenever Indigenous Peoples began to experiment with controlled burns—not just in North America but in Africa, Australia, and Asia, too—they did so within the context of existing fire regimes that had developed over many millions of years. They learned to burn from the greatest teacher of all, the original Prometheus: our living Earth. Over millennia, humans became the coconductors of fire's ecological rhythms. Eventually, we would alter them more drastically than any creature before us—sometimes to marvelous effect, sometimes with dreadful consequences.

THE DAY AFTER MEETING Frank Lake at his property, I ventured northeast of Orleans, past Somes Bar, and into Klamath National Forest, near an area known as Rogers Creek. Moss pillowed every rock, trunk, and stump. Wisps of pale lichen hung along the length of every branch, as though the trees were antique chandeliers caked in melted wax. Persistent fog and a light, intermittent rain evoked the atmosphere of a cloud forest. The stout smell of wet earth and rotting leaves flavored the air, muddled with their near opposites: the scent of woodsmoke and ash.

Dozens of people dressed in flame-resistant clothing—mustard yellow shirts and pine green pants—paused along a forest service road to adjust their hard hats, strap propane tanks onto their backs, and test the torches connected to them: long, thin, metal rods with a stream of flaming gas at one end. Although they were all certified firefighters, they were not there to extinguish anything. They had come to burn. A diverse group of foresters, conservationists, paramedics, members of local Indigenous communities, students, and pyrophiles, they had traveled from near and far to participate in a program known as TREX: prescribed fire training exchange. Founded in 2008 by the U.S. Forest Service and The Nature Conservancy, TREX teaches people how to use controlled burns to benefit ecosystems and reduce the chance of severe wildfires.

The firefighters—some of whom prefer to be called fire*lighters*—moved carefully down steep slopes into the midst of the forest, searching for large piles of branches and brush, which crews of foresters had cut and stacked in the preceding months, covering their centers with wax paper to keep them dry. When a firelighter found a brush pile, they would push their torch into its heart and squeeze a lever or turn a knob to increase the flow of gas, scorching the pile's interior with a fierce orange flame.

At first, it seemed like some of the piles were too wet to burn properly. Although they spewed plumes of smoke like volcanoes stirring from slumber, they did not erupt in flame. A little rain is beneficial for pile burning, as it prevents fires from becoming too big and hot, but too much moisture defeats the purpose. Tossing branches onto an especially large mound that was sustaining a sizable fire, forest ecologist and firefighter Michael Hentz explained that the piles needed time to burn and dry from the inside out before catching fire in their entirety. As the day progressed, more and more piles began to burn, sometimes so vigorously that they lofted ash and embers high above us. Soon the whole forest seemed to glow and crackle within shifting layers of fog and smoke. Although I knew these fires were intentional, the sight of them still provoked some deeply embedded survival instinct—a stubborn feeling that something was wrong. It was strange to see the forest on fire. It was beautiful, too. Surveying the many heaps and rings of wood with flames leaping from their centers, it felt like we had stumbled into a colony of phoenix nests.

"This is one of the most important steps in reintroducing fire back to this mountainside," said Zack Taylor, burn boss and one of the key organizers of the day's events. His face was framed by a short brown beard, thin-rimmed glasses, and a baseball cap; a radio transceiver was clipped to the inside of his pocket. The fifty acres on which they were burning, he explained, were populated with a mix of tanoak, black oak, canyon live oak, big leaf maple, madrone, and a surfeit of spindly Douglas firs. "The ecological trajectory we want is one in which we have less conifers and more healthy hardwoods," he continued, occasionally pausing to spit. "They're an important cultural food source, and they have a lot of value for wildlife, but they're lacking on the landscape because of a hundred years of fire exclusion. We're not trying to go back in time. I don't think that's really possible. All we can do is say, 'Here's what we would like. What are the logical steps to get us in that direction?' And primarily, that would be the use of fire."

Eventually, when this part of the forest has been sufficiently thinned, Taylor and his colleagues plan to return and perform a broadcast burn, a typically low-level fire that sweeps across a predefined area. In the process, it will consume all the dry, combustible plant matter on the forest floor, as well as the remaining shrubs and scraggy growth, without killing large, established trees. Teams of firelighters will most likely conduct the broadcast burn with driptorches, canisters that dribble a flaming mix of gasoline and diesel onto the ground.

Fire ecologist Scott Stephens has witnessed the effects of such burning firsthand. "Fire has a profound ability to provide essential feedback within a living system," he says. "There are some places in Yosemite where we've allowed lightning fires to burn without suppression for fifty years. You get big, old trees in nice open conditions interspersed with patches of shrubs, burned logs, dead trees that are still standing, and lots of regeneration in between. Some people might look at that and think it's a mess because they are used to wall-to-wall forest, but it's the closest thing we have to a functioning fire regime in California. In 90 percent of cases, fires that happen there go out on their own. It's completely self-regulating."

When Frank Lake was a boy, TREX was many decades away from existing, Indigenous burning traditions were often prohibited by law, and prescribed burns in the West were uncommon. He remembers some of his family members and elders lamenting the loss of fire on the landscape. His father, Bobby Lake-Thom, also known as Medicine Grizzly Bear, and his grandfather, Charlie "Red Hawk" Thom, were both Karuk medicine men. They would take him to sacred sites called medicine places, where they would pray and occasionally make small fires. Sometimes they visited areas where someone had defied the law by performing a controlled burn to clear space for huckleberries, improve acorn harvests, or encourage fresh hazel growth for basket weaving. "My grandfather Charlie, who was a ceremonial leader, talked about the importance of fire to our culture and how since the Forest Service took that away the Earth was getting sick and dying," Lake told me during one of our conversations. "You can't always wait for lightning to strike. As a fire-dependent culture, you have to get out and burn."

Lake's parents divorced when he was about five, after which he divided his time between homes in Orleans, Eureka, and the Yurok Reservation. From middle school on, he lived with his mom and stepdad in Sacramento. "I was not the strongest student," he remembers. "I was the kind of kid who always took off and went and played down in the forest." He was conditionally accepted to the University of California, Davis, where he had to take a remedial English class. "I had my cultural teachings, but I didn't have the Western academic skill set," he says, "and that was a problem for me because I wasn't a strong writer. I didn't even pass my first wildlife ecology class. I knew all the species, knew all their habitats. I could identify them by track, by skull, by fur and feather—but I couldn't spell their Latin names."

After graduating from UC Davis in 1995, Lake spent a few years in southern Oregon and northern California working as a fisheries biologist for the U.S. Forest Service. Then, in 1999, the Megram Fire devastated the part of the world he called home, burning more than 125,000 acres of national forest, Indian reservation, and private land. At the time, it was one of the largest wildfires in California's history. Thick smoke filled the sky for weeks. In steep areas where most of the trees and vegetation had been incinerated, cliffs eroded and landslides clogged rivers with sediment. For Lake, the crisis precipitated an epiphany: the Megram Fire was not the sole purview of his colleagues in fire ecology and forestry; it was directly relevant to his work as a fisheries biologist, too. Trees and fish, fire and water—they were all connected. "The foresters and the fire folks were looking up at the ridge in the forest, and the hydrologists and fisheries people were looking at the creeks and rivers," he says. "We were supposed to be managing all these resources, yet we couldn't seem to turn around and encompass a broader perspective. And I thought, 'What's the one thing that unifies the ridge to the river? Fire.' "

Lake remembered how the elders in his community had taught him that too little and too much fire were equally harmful for fish and other aquatic species. If fires were too small and infrequent, trees would crowd and parch the landscape, preventing rainfall from refilling springs and reservoirs. If fires were too big and destructive, there would not be sufficient vegetation to sponge up excess water or roots to hold the soil in place, resulting in more floods and landslides. In the aftermath of the Megram Fire, these ecological connections seemed to glow with new significance. Realizing how much more there was to learn, Lake decided to leave his job and return to school.

In the fall of 2000, he began a PhD in environmental science at Oregon State University, where he was one of very few Indigenous students. One afternoon, he stood up in class and challenged a white professor who had belittled the importance of controlled burns by Native Peoples in shaping the North American landscape—a prevailing attitude at the time. "You know, Professor, I think you're giving a biased perspective," Lake remembers saying. "I don't know about that," the professor replied. "Well, I don't think you've done a thorough literature review," Lake said. "Okay," the professor said. "If you can substantiate your claim and document the evidence, I'll consider it."

Two days later, after exhaustive research in the library, Lake returned to class with dozens of handouts documenting Indigenous burning traditions. "Here's what you discounted," Lake told his professor. "You said Native People didn't have a reason to burn their environment. Actually, they had quite a few reasons. You left out cultural anthropology and archaeology and oral histories of the very system you're supposed to be an expert for. I see you interjecting in your lecture a bias that is counter to accepting and acknowledging Indigenous Peoples. And I'm calling you out on it."

That professor was on Lake's original dissertation committee, and, as Lake remembers it, "he didn't like having his authority or perspective challenged." In 2007, several years after petitioning for a more diverse committee, Lake successfully defended his thesis on integrating Indigenous ecological knowledge and Western science to reintroduce prescribed fire to northwestern California, with a particular focus on using controlled burns to manage sandbar willow for basket weaving.

Shortly thereafter, he transitioned to a full-time job as a research ecologist with the U.S. Forest Service.

Since then, he has published many research papers on prescribed burns and Indigenous stewardship of natural resources. In 2018, nearly two decades after the Megram Fire and after years of fighting anti-Indigenous biases in academia, he finally published a study demonstrating a clear link between fire and fish. Lake had learned from his elders that the Karuk sometimes used controlled burns to "call back the salmon from the ocean," which many of his academic peers dismissed as folklore. Using NASA satellite imagery and meteorological records, however, Lake and two colleagues demonstrated that by reflecting heat and light, wildfire smoke lowered the temperature of rivers, improving the survival of migrating salmon and other cold water–adapted species, especially during heat waves—one of the many nuances of fire ecology that Native Peoples discovered thousands of years before it was a formal scientific discipline.

Lake has also become a key figure in collaborations between the Forest Service and Indigenous Tribes, as well as a champion of the growing movement to return fire to western North America. Thanks in large part to advocacy by Lake and other Indigenous leaders, both federal and state government agencies are increasingly open to using prescribed fire to reinvigorate ecosystems and reduce the likelihood of more disastrous megafires. In January 2022, the Forest Service announced a new national strategy for confronting the wildfire crisis, calling for "a paradigm shift" in land management policies. The plan promises to collaborate with "States, Tribes, local communities, private landowners, and other stakeholders" to increase pruning and controlled burns by up to four times current levels, supported by three billion dollars from the Infrastructure Investment and Jobs Act passed in 2021. "We need to thin western forests and return low-intensity fire to western landscapes in the form of both prescribed and natural fire," the report said, "working to ensure that forest lands and communities are resilient in the face of the wildland fire that fire-adapted landscapes need."

I asked Lake what he envisions for the future. "I want to scale up," he said, with typical fervor (he once described himself to me as "kind of an intense person"). "If my gold standard is my half-acre orchard, we should have fifty thousand acres of it, because that's what we need around here. I have learned this Western system of sound, credible science. I've been able to use that to demonstrate that Indigenous practices can fulfill desired objectives for carbon sequestration, climate resilience, food security, biodiversity, and the mitigation of severe wildfires. The Tribes shouldn't have to acquiesce their stewardship to government agencies. They should be co-leaders, co-managers. We're starting to do it. What I do is no longer questioned the way it was before. You serve by example, and that is replicated and carried out in its own unique way in other places."

His reply reminded me of a conversation I had with Margo Robbins. "Fire belongs in the hands of the people, not just government agencies," she told me. "Fire is meant to be part of the ecosystem. It's our responsibility as human beings to learn how to use fire so that we can assume our proper role in putting fire on the land. And I don't mean just Native People—I mean everyone."

WHEN FIRE FIRST BECAME part of the Earth system, it was highly volatile. The rhythms that characterize modern fire-adapted ecosystems took hundreds of millions of years to form. Earth's earliest wildfires were probably fitful and erratic, flickering among the amphibious flora of fens and bogs more than 400 million years ago. In contrast, during the Carboniferous, between 375 and 275 million years ago—when atmospheric oxygen levels were at their peak and giant dragonflies soared through the air—fires were frequent and rampant, burning as hot as 752 to 1,112°F and incinerating even lush, wet vegetation. For a long time, oxygen levels, and the frequency and intensity of wildfires, continued to fluctuate widely.

Around 200 million years ago, however, something appears to have changed: the amount of oxygen in Earth's atmosphere began to stabilize, remaining within the relatively narrow window of 20 to 30 percent. Innovative experiments by University of Exeter Earth scientist Claire Belcher have demonstrated that fires cannot sustain themselves if the atmosphere contains less than 16 percent oxygen; conversely, if oxygen exceeds 23 percent, wildfires are much more likely to blaze out of control and essentially anything that isn't drenched or submersed in water becomes flammable. In the past 55 million years or so, the level of atmospheric oxygen has been more stable than ever, hovering right around 21 percent, which is high enough to support occasional wildfires and an incredible diversity of complex, fire-adapted life, yet not so high that any stray spark will ignite an unstoppable inferno. Scientists have long struggled to explain this remarkable equilibrium. In the past couple decades, however, they have begun converging on a possible answer: the coevolution of fire and life.

Geoscientist Lee Kump was one of the first scientists to formally publish a theory of this particular planetary balancing act, which was further developed by Tim Lenton, among other researchers. The key to understanding their model is a chemical element whose name is synonymous with starlight: phosphorus. All living organisms require phosphorus, which is an essential building block of DNA and cell membranes, but natural sources are limited. Most phosphorus is trapped in rock and is gradually liberated by rain, ice, and wind. When microbes, fungi, and plants populated land surfaces and began to recycle water and break apart the planet's crust with roots and acids, they expedited the release of phosphorus from rock and increased the flow of phosphorus from land to sea via rivers. This, in turn, enhanced the productivity of both terrestrial plants and ocean-dwelling photosynthesizers, such as phytoplankton. That elemental link between land and sea, Kump and his colleagues propose, ultimately became the foundation for a vital feedback loop.

When the level of atmospheric oxygen rises too high, fires become rampant and destroy immense tracts of vegetation, reducing the overall capacity of land plants to liberate and capture phosphorus. At the same time, more of the phosphorus that is available ends up in the ocean, where it is used by plankton and algae. Marine photosynthesizers do not use that phosphorus as efficiently as their terrestrial counterparts, however. For every atom of phosphorus land plants acquire, they are able to store one thousand atoms of carbon in their bodies; in contrast, ocean plants store only one hundred. Thus, higher oxygen levels and raging wildfires ultimately impair the total productivity of Earth's photosynthetic organisms and decrease the amount of organic matter that is buried in bogs and ocean sediments, weakening the very mechanism by which oxygen accumulates in the atmosphere. Over millions of years, the level of oxygen drops, raging fires seethe and sputter, and land plants recover. Although this theory is not yet textbook science, a growing cadre of scientists think the feedback loop it describes has stabilized the amount of oxygen in Earth's atmosphere for at least 50 million years.

When the Gaia hypothesis rose to prominence in the 1980s, some of its most controversial tenets were that life controls global climate in order to benefit itself and that the Earth system as a whole actively "seeks an optimal physical and chemical environment for life on this planet," as James Lovelock and Lynn Margulis phrased it early on. As Earth history shows us, that is not quite true.* To the contrary, many forms of life—from microbes to trees to bipedal apes—have caused or exacerbated some of the worst crises in Earth history. And there is no single "optimal" state of the planet that would suit all of the myriad and wildly diverse types of life that have existed in the past four billion years. In general, though, if given enough time and opportunity, life and environment seem to coevolve relationships and rhythms that ensure their mutual persistence. There is nothing teleological about this. Such persistence is not designed or planned. It is the outcome of ineluctable physical processes that are distinct from, but related to, the processes that govern the evolution of species.

* "I made mistakes," Lovelock wrote in updated editions of *Gaia*. "Some were serious, such as the idea that the Earth was kept comfortable by and for its inhabitants, the living organisms. I failed to make clear that it was not the biosphere alone that did the regulating but the whole thing, life, the air, the oceans, and the rocks. The entire surface of the Earth including life is a self-regulating entity and this is what I mean by Gaia."

All complex multicellular organisms have evolved numerous ways to maintain homeostasis—to preserve a steady state of physical and chemical conditions essential to their continued existence. All complex organisms are also chimeras: their genomes are patchworks stitched with genes introduced by viruses and pilfered from other species; some of the organelles in their cells were once free-living bacteria subsumed in the emergence of multicellular life; their bark, fur, or skin teems with trillions of microbes, competing, cooperating, and multiplying in secret societies. Any individual plant, fungus, or animal is, in effect, an ecosystem. If such composite creatures can evolve homeostasis—a point about which there is absolutely no disagreement—then perhaps an analogous phenomenon, which science does not yet fully understand, occurs at the scale of forests, grasslands, coral reefs, and other ecosystems.

Ecosystems may not compete and reproduce the way organisms and species do, but some scholars have proposed that they should be regarded as living entities capable of self-regulation and evolution. The coevolution of the organisms and habitats that compose a given ecosystem influences how that system changes over time. An ecosystem, then, does not evolve passively; it effectively changes itself through inevitable feedback loops—at least to an extent. Although the particular species and habitats within these systems shift dramatically over time, the fundamental relationships that define them, the cycles and webs that bind prey and predator, flower and bee, leaf and flame, and the physical infrastructure that life creates—the rich soils, webs of root and fungi, reefs, and ocean sediments—typically persist or, if they are demolished, regenerate in some form. Networks of species that happen to help sustain the system as a whole will be favored, whereas those that undermine the system to the point of collapse will ultimately eliminate themselves, even if they profit in the short term. The most resilient ecosystems—those best able to adapt to challenges and crises—will survive the longest. Perhaps this phenomenon of persistence extends to the planet as a whole. Not an intention to persist but a tendency; not an imperative but an inclination.

Whether cell or cetacean, prairie or planet, living systems find ways to endure.*

SHORTLY AFTER FRANK LAKE and I toured his property, he drove us into Six Rivers National Forest to see more of the local landscape. While navigating a steep, curving road, we encountered a plume of smoke rising gently from within a tangle of ferns and blackberry vines. Lake slowed down so that we could get a better look. Behind the smoke, a river of ash snaked between the charred remains of several trees, now hollowed and tapered to a peak, like small, ancient temples.

"Should we stop here?" Lake said. "We can pull around the side. This could be a carryover from a prescribed burn."

We got out of the car and explored the area. Just across the road, curtains of smoke, almost as thin and diaphanous as steam, drifted between blackened Douglas firs, mossy oaks, wild roses, and golden-leaved maple trees. Several downed trees had been reduced to brittle, coal-dark stubs. Near the center of the scene a pool of gray ash spilled from an especially large log, which was itself black and crumbling, like a partly cremated canoe. Someone had clearly burned piles here a few days earlier, Lake explained, perhaps TREX participants or other local firelighters.

"Due to fire exclusion, this area has grown in," Lake said. "You have an increase in falling trees and logs that build up, and a region that used to have more fire and nutrient-cycling now has more decomposition."

Nearby, Lake pushed some forest litter, ash, and soil to the side and showed me how the fire had found a buried log, which was still smoldering underground—the source of some of the smoke. "Even though we've had rain, there are still dry logs underneath here," he said. "Burns like this are going to create diversity." When the log burned up com-

* Although the ideas discussed in the preceding three paragraphs have long been controversial within scientific circles, they continue to attract serious interest from mainstream scholars. To learn more, see the sources for this chapter listed at the end of the book, especially those by Bouchard, Doolittle, Dussault, and Lenton.

pletely, he explained, it would open space in the soil through which air and water could move more freely. "That will be a macropore," he said. "There will be a moisture site in there. This will be a place for critters. The idea is to increase the persistence of things that are most valued and to do that through a lens of human responsibility. Not only reducing wildfire threats to these communities like we did here, but also showing our responsibility to the forest to improve its overall resilience."

Later that day, after picking up a deer that Lake had hunted and dropped off at the general store in Happy Camp for processing, we got to talking about his revived orchard. Lake told me how he speaks to his oak trees, explaining his intent, and how he often begins prescribed burns with a prayer. "A while back, I was looking for a little smudge stick of sage and cedar to start to pray with, to kind of clean things and be right," he said. "I was digging around in my shop and I found one of the last sage bundles that my dad made before he died. And when I pulled it out"—he paused, his voice cracking—"it smelled like my dad."

"He'd say, 'When you're going to do something good and you need your spiritual grounding and protection, burn this and pray for that ability to do those good things.' I burned that. BIC lighter, smudged myself up, and then from there I lit up my propane wand and burned my area. It was done with prayer. It was done with good intent. It was done not in fear but reverence for my trees. It was done with the fulfillment to those trees that I talk to and pray to and sing to."

"The trees can't do it by themselves," he said. "The forests can't do it by themselves. We can't do it by ourselves. When are we going to see that we're a part of a mutualistic process of climate adaptation and resilience? When are we going to accept that the only way to survive is together?"

CHAPTER

9

WINDS OF CHANGE

AS A YOUNG CHILD, YI GUO RARELY SAW A BLUE SKY. WHENEVER SHE looked up, her eyes usually met a thick gray haze. Coal-mining was the dominant industry in her hometown of Tongchuan, China. Dust from the mines and pollution from nearby cement factories continually choked the air. By the 1980s, the smog was so dense and persistent that satellites could no longer reliably photograph Tongchuan, earning it the nickname of "invisible city."

Just about everyone Guo knew worked in coal in one way or another. Her grandfather on her mother's side suffered a spinal injury during a mining accident that left him leaning on a walking stick for the rest of his life. Her paternal grandfather developed chronic lung disease. Guo remembers a period in which her father tallied accidents and deaths in the mines on a near daily basis. When her mother went clothes shopping, she primarily chose dark fabrics, because in the polluted air of Tongchuan, light colors soiled too quickly.

Around the age of nine, Guo moved to Xi'an, one of the ancient capitals of China, where she eventually went to college. She excelled in math and science and chose to study mechanical engineering. After earning her master's degree, she moved to the United States to continue her graduate studies at the Ohio State University. There, one of her professors asked her to help figure out why a wind turbine was malfunctioning. Guo had been vaguely aware of renewable energy before that point—she remembers marveling at hundreds of small wind turbines in a grassy field during a trip to Inner Mongolia—but she had never studied it in detail.

Renewable energy is at once abundant and precious. Earth is a giant, part-molten rock, radiating heat from within, wrapped in ribbons of air, veined with rivers, sloshed with ocean, awash in sunlight, and lush with self-regenerating vegetation. Because these resources are readily available, continuously replenished, and essentially inexhaustible, the energy we derive from them is itself renewable. In contrast, the reserves of coal, oil, and gas on our planet, while vast, are finite and concealed, requiring intensive extraction and transport, often at the expense of both human and environmental health. Moreover, the carbon dioxide released by the combustion of fossil fuels thickens the atmosphere's heat-trapping cloak of greenhouse gases,* raising global temperature, weirding weather, and creating a far more unpredictable and inhospitable climate.

Guo's research on wind turbines immediately fascinated her. She loved learning about these extraordinary machines that had already evolved so rapidly, yet still had so much potential. She and her colleagues eventually discovered several design flaws in the wind turbine they had been asked to investigate and devised a solution to the problem. Working on wind power was a stark contrast to her family's experiences in the coal mines—a completely different way of engaging with the planet's resources and meeting humanity's energy needs. The more Guo reflected on that distinction and on the importance of clean energy for the future of human civilization, the more inspired she became.

She is now a professor of mechanical engineering at the Technical University of Denmark, where she focuses on the design and manufacture of wind turbines. "I don't see how it's beneficial for anyone in the long run to keep using fossil fuels," she says. "We need to use what we already get from the wind, the sun, and the earth, instead of disrupting

* A greenhouse maintains warmth by preventing the sun-heated air within it from dispersing. Methane, carbon dioxide, water vapor, and other so-called greenhouse gases work in an analogous but distinct way, permitting sunlight to pass through them and heat Earth's surface but impeding a portion of the thermal radiation that tries to escape to space.

our planet. Otherwise, what will be left for our kids? For our grandkids? For all the new generations of human beings?"

THE STORY OF HOW our species altered Earth's climate begins much earlier than is often believed, long before the Industrial Age. When we split from our last common ancestor with chimpanzees somewhere between five and nine million years ago, we inherited a world made and remade over eons by earlier forms of life. The fertile soil, lush forests, bountiful oceans, blue sky, and breathable air were gifts bequeathed to our ancestors by nonhuman predecessors. So, too, was the possibility of further change: the opportunity to discover new resources and new ways of living.

In Earth's early history, the only sources of energy widely available to living organisms were sunlight, the planet's inner heat, and the byproducts of spontaneous chemical reactions between water and rock. Primordial microbes initially evolved to use these types of energy and, later, to consume one another. In turn, algae, plants, and the oxygen they exhaled became essential fuels for new waves of complex animal life. An abundance of land plants in a highly oxygenated atmosphere also ignited a novel source of light and heat: fire.

When our ancestors harnessed fire, they transcended the energetic constraints imposed on all other animals. Instead of eating only raw vegetation and flesh, early humans began to cook their food, making it more digestible and extracting more of its calories. This enriched diet ultimately permitted the evolution of far larger and denser brains, supporting the suite of cognitive abilities that has made our species so successful. Fire is only as powerful as that which feeds it, however, and for most of human history, our ancestors only knew how to burn a single and rather inefficient type of fuel: living or recently living plants, whether in the form of leaves, wood, hay, or mastodon manure.

That changed with the discovery of fossil fuels, which are energy-dense deposits of ancient life compressed and cooked deep within the planet's crust—hence the "fossil" in their name. Earth's coal deposits primarily formed in hot, humid swamps and wetlands more than 300 million years ago, during the geologic period to which they lend their name, the Carboniferous (from the Latin for "coal-bearing"). When massive ferns, scaly-trunked *Lepidodendrons*, and giant relatives of horsetails died, they were sometimes buried by water and sediment before microbes fully decomposed them. As layers of dead vegetation accumulated, they were subjected to intense heat and pressure. Over millions of years, those forces rearranged the entombed plants at the molecular level, breaking apart existing compounds and forming new ones, turning the jungles of primeval Earth into peat and eventually into coal. In contrast, natural gas and petroleum, or crude oil, are mostly composed of algae, plankton, and other aquatic life subjected to extreme pressures and temperatures at the bottom of lakes and on the seafloor during the more recent Mesozoic Era (252 to 66 million years ago).

Whereas fire was a universal element of early human cultures, the adoption of fossil fuels was much more staggered and piecemeal. In the Bronze Age, between 2200 and 1900 B.C., people in what are now Inner Mongolia and China's Shanxi province excavated coal from shallow deposits and burned it for heat, especially when wood was scant. Ancient Romans and medieval Europeans also relied on coal for warmth and for smelting iron ore. By 60 B.C., and possibly much earlier, the Chinese were drilling for oil and natural gas, eventually learning to channel the fuels through bamboo pipelines and burn them beneath iron pans to evaporate brine and yield salt. The ancient Chinese and Arabs also burned oil and gas for light and heat.

In the sixteenth century, as forests dwindled due to overharvesting, England began extracting massive quantities of coal from abundant and relatively accessible deposits. Burning a lump of coal provided significantly more energy than burning the same amount of wood or vegetation. By the seventeenth century, coal powered the majority of England's industries and heated most of its homes. Eventually, many other countries began using coal extensively, too. "Every basket is power and civilization," wrote Ralph Waldo Emerson. "For coal is a portable climate. It carries the heat of the tropics to Labrador and the polar circle; and it is the means of transporting itself whithersoever it is wanted."

Between the late eighteenth and mid-nineteenth centuries, fossil fuels sustained one of the most important technological and socio-economic transitions in history: the Industrial Revolution, a period of rapid innovation in manufacturing technology in Europe, North America, and Asia. Coal-burning steam engines were initially developed to pump water out of frequently flooding coal mines. As their efficiency improved, steam engines began to power spindles, looms, mills, factories, ships, and locomotives. A series of coal-dependent breakthroughs in metallurgy resulted in more affordable metals of higher quality, which further stimulated the production of new machines. Expanding networks of canals, roads, and railways allowed people to efficiently transport large volumes of food and fuel over great distances. Cities began illuminating streets with gas lamps as rural areas swapped whale oil for kerosene, both of which would eventually be displaced by electric lights. The commercialization of the internal combustion engine in the late nineteenth century and the subsequent mass production of motor vehicles drove demand for petroleum. Around the same time, the introduction of the modern steam turbine and its adoption in power plants started to make electricity much more widely available.

In the 1890s, coal surpassed wood as the most widely used fuel in the world and continued to dominate throughout most of the twentieth century. By the turn of the twenty-first century, however, oil and natural gas had grown from only 4 percent to 64 percent of the global energy supply, surpassing coal; in addition to being more energy-dense, they were often easier and cheaper to store and transport. At the time of writing, fossil fuels continue to supply about 80 percent of global energy. Transportation, manufacturing, and heating are especially dependent on fossil fuels, as are the production of iron, cement, fertilizers, and electricity. Because electricity itself seems so pure and ethereal, it's easy to forget that the majority of the electricity coursing through the modern world—around 64 percent as of 2019—is generated by burning fossil fuels in power plants to spin steam turbines. Despite all the innovation and progress of the past three centuries, the global economy still runs on an elaboration of what is fundamentally Industrial Age technology.

When our ancestors first stumbled upon coal, oil, and natural gas so long ago, they did not understand the origin or composition of these strange substances. But we do. We've known for more than a century that fossil fuels are combustible crypts containing the power of countless deceased life forms that collectively absorbed hundreds of millions of years of sunlight. Environmental scientist and policy analyst Vaclav Smil has calculated that a single gallon of gasoline represents *one hundred tons* of ancient life, roughly equal to twenty adult elephants. Every sedan with a typical fifteen-gallon gas tank demands the equivalent of three hundred elephants simply to keep running. Fossil fuels are not just conveniently concentrated forms of energy—they are outrageously extravagant. A fossil fuel is essentially an ecosystem in an urn.

Humans began increasing the flow of greenhouse gases to the atmosphere long before the Industrial Revolution, primarily by devastating entire ecosystems: by hunting megafauna to extinction, clearcutting forests, and replacing native habitat with methane-wafting rice paddies and herds of livestock. When the first industrial nations began unearthing and burning fossil fuels en masse, however, they initiated an unprecedented distortion of Earth's carbon cycle. In 1750, annual worldwide CO_2 emissions from human activity were an estimated nine million metric tons. A century later, they had increased more than twenty times to around 197 million tons. By 1950, they had grown another thirty times, reaching six billion tons. In 2021, human activity generated more than 36 billion tons of CO_2—the highest level in history.* Humans now emit between 60 and 120 times more carbon dioxide each year than all the world's volcanoes combined.

Since preindustrial times, human activity has released close to 2.5 trillion tons of carbon dioxide to the atmosphere. That invisible mass

* Remember that one ton of carbon = 3.67 tons of CO_2. Thus, humanity's annual global carbon emissions are 36 billion tons of CO_2 or 9.8 billion tons of carbon. If we account for all greenhouse gases emitted through human activity—not just CO_2 but also methane, nitrous oxide, and fluorinated gases—annual emissions are about 50 billion tons of CO_2e, or CO_2 equivalent.

of CO_2 suspended in the sky is more than twice as heavy as the entire collection of living creatures on the planet and almost double the weight of everything made by humans that is still in use: all the metal, concrete, glass, and plastic—all the cities, roads, factories, dams, jumbo jets, air fryers, gas-powered leaf blowers, and motorized ice-cream-cone spinners. The United States alone has emitted around 25 percent of the total, twice as much as the second-biggest contributor, China, at 12.7 percent.* Together, North America and Europe are responsible for 62 percent of historical emissions. Today, the wealthiest half of the world's countries continues to produce 86 percent of all CO_2 emissions.

Although the ocean and continents, and the life they harbor, have absorbed a large portion of humanity's excreted carbon, much of it has remained in the air, increasing the atmospheric concentration of CO_2 by 50 percent, from about 277 parts per million (ppm) in 1750 to around 420 ppm today. Some research indicates that the last time there was this much carbon dioxide in the atmosphere was about four million years ago, during the Pliocene, when the average global temperature was 5.4°F hotter than today, sea level was up to 82 feet higher, and vast forests grew in what is now treeless Arctic tundra.

Though the level of CO_2 in the atmosphere has fluctuated dramatically throughout Earth's history, most of these shifts were relatively gradual, spanning anywhere from tens of thousands to millions of years. When carbon rapidly floods the atmosphere, terrible things happen. About 56 million years ago, during a major climatic crisis called the Paleocene–Eocene Thermal Maximum, roughly three to seven trillion tons of carbon were released into the atmosphere, perhaps due to extreme volcanic activity, increasing the global average temperature by 9 to 14.4°F, heating and acidifying the oceans, and driving many deep-sea species to extinction. The displacement of carbon underlying this calamity happened over many thousands of years. Humanity is

* Russia, Germany, the United Kingdom, Japan, India, France, Canada, and Ukraine round out the top ten. These rankings are primarily based on emissions from burning fossil fuels. When emissions from deforestation and other large-scale environmental modification are included, Brazil and Indonesia move into the top ten but do not displace the United States, China, or Russia.

releasing a comparable amount of carbon in just a few *hundred* years. Although it is impossible to say for certain, especially given all that is unknown about the planet's earliest chapters, it is plausible that throughout the greater part of Earth's 4.5-billion-year history, this much carbon has *never* been released to the atmosphere this quickly.

From a geological perspective, the past twelve thousand years, from the early stages of agriculture onward, have been a period of remarkable climatic stability—an especially harmonious phase of Earth's life story. Greenhouse gas emissions are ushering in an era of what would be, in comparison, utter cacophony. Collectively, emissions from the combustion of fossil fuels, deforestation, habitat destruction, farming, refrigeration, and other human activities have increased Earth's average surface temperature by about 1.2°C (2.16°F) relative to the late nineteenth century. That might seem insignificant, but raising the average temperature of the entire world by even one degree requires an immense amount of energy. Climate models suggest that if humans everywhere instantaneously ceased all CO_2 emissions, global temperature would stabilize fairly quickly but would not return to the preindustrial baseline for hundreds to thousands of years. The consequences of this massive perturbation of the Earth system have already been manifold and severe.

As glaciers retreat, permafrost thaws, and ice sheets melt, Earth is losing some of its most reflective surfaces, decreasing its albedo. The lower Earth's reflectivity, the hotter it gets, which melts even more ice. Due to this planetwide thaw and the fact that water expands as it warms, average sea level has risen nearly nine inches since 1880. Regardless of energy policy going forward, the amount of greenhouse gases already in the atmosphere dictates that sea level will continue to rise by several feet over the next few centuries, eroding shorelines, inundating wetlands, intensifying storm surges, and endangering coastal communities and low-lying island nations around the world.

As Earth heats up, the atmosphere's capacity to hold water vapor increases. On a hotter, wetter planet, floods and storms become more powerful and, depending on the location, more frequent. In 2022, following extreme heat and monsoon rains, Pakistan suffered the worst floods in its history, killing more than 1,500 people and displacing

30 million. In the fall of 2023, Storm Daniel caused devastating flooding in Libya, Turkey, Greece, and Bulgaria, collapsing dams, killing at least several thousand people, and leaving more than ten thousand missing.

Heat waves and wildfires in the past decade have continually broken one record after another. In the summer of 2021, the Pacific Northwest endured extreme heat unlike anything in its history. The air temperature in some parts of Portland, Oregon, rose to 124°F, while pavement reached a scalding 180°F. Roads buckled and cables warped, forcing the city to suspend its streetcar and light rail services. In late June of the same year, the village of Lytton in British Columbia reached a staggering 121.3°F—a new national record for Canada that also surpassed any temperature ever recorded in Europe and South America at the time. The next day, a wildfire all but destroyed the village. The following year, a series of brutal heat waves killed more than twenty thousand people in Europe. In 2023, North America experienced the worst wildfire season in its recorded history; as of late September, more than 43 million acres had burned in Canada—about ten times the number of acres burned in 2022, and around 5 percent of the country's entire forested area. In August of the same year, a firestorm reduced Lahaina, Maui, to ash, killing close to one hundred people—one of the worst natural disasters in Hawaii's history and one of the deadliest wildfires in the United States since the early 1900s.

The combination of heat and humidity in some equatorial countries is now so extreme that the human body has no adequate physiological defenses against it. At the same time, rising temperatures make dry places even drier, expanding deserts and increasing the severity and frequency of droughts. In arid regions, access to clean drinking water will become even more fraught than it is today, inciting new conflicts. Crops in some northern climes may benefit from the surplus heat and CO_2, but global crop yields are predicted to decline due to extreme weather, degraded soils, and more pervasive pests.

Fossil fuel emissions and other forms of atmospheric pollution have already taken a massive toll on public health, in part by increasing rates of respiratory illness—a problem exacerbated by wildfire smoke and dust from impoverished soils. Meanwhile, tropical diseases and the pathogens that spread them are expanding their range, and pandemics of novel illnesses like COVID-19 are becoming more common and more dangerous.

Diverse species are already migrating to higher latitudes and altitudes to find cooler weather. Flowers are blooming earlier than in the past, falling out of sync with their coevolved pollinators. Forests and grasslands are shrinking, drying out, burning, and struggling to adapt. As the ocean absorbs heat and CO_2, it is losing its ability to buffer climate change. Some scientists have predicted that by the end of the century if not earlier, most of the world's once kaleidoscopic warm-water coral reefs will be reduced to meager fragments.

The enormity of this turmoil is heightened by one of the most terrifying aspects of the current planetary crisis: its unpredictability. Some aspects of the Earth system, including critical inflection points and feedback loops, are so complex and difficult to model that scientists cannot reliably anticipate their outcomes. The physics of water vapor and clouds, for example—to what degree their simultaneous heating and cooling of the planet modulate overall climate change—are especially tricky to simulate, as are fluctuations in ocean currents and jet streams with potentially catastrophic repercussions. Equally horrifying is the speed with which this is all happening. Earth has recovered from disasters many times before, including some that were, in the grand scheme, much more devastating than modern climate change. But in every case, the living planet required tens of thousands to millions of years to restabilize. The world that emerged from each of those calamities was often radically different from the previous version, with entirely new forms of life replacing those that went extinct. We cannot rely on Earth's thermostat and other innate self-stabilizing processes to spare us from the current planetary crisis because they operate on geological timescales irrelevant to the year-to-year survival of individual species and civilizations.

If humanity does not drastically cut greenhouse gas emissions, Earth will become a planet incapable of supporting the world as we have known it: the world that our species evolved in and the one we have been constructing since we started using tools and making fire.

Many of the ecosystems and much of the infrastructure on which modern human societies depend will collapse. As highly adaptable and tenacious creatures, humans are unlikely to go extinct because of climate change alone, but hundreds of millions to billions of people—especially those who live in vulnerable communities with the greatest environmental risks, the fewest resources, and the least capacity to adapt—will suffer the ravages of extreme weather, lose their homes and livelihoods, or die from starvation, disease, heat stress, storms, and flooding. Countless nonhuman species will disappear, breaking vital biogeochemical cycles and depleting Earth of its diversity, vibrancy, and beauty. "The cumulative scientific evidence is unequivocal," the Intergovernmental Panel on Climate Change wrote in its Sixth Assessment Report in 2022. "Climate change is a threat to human well-being and planetary health." Any further delay in concerted global action to both mitigate the current crisis and adapt to its unavoidable consequences will "miss a brief and rapidly closing window of opportunity to secure a liveable and sustainable future for all."

ABOUT 15 MILES SOUTHEAST of Reykjavik, in the shadow of an active volcano known as Hengill, stands the Hellisheiði geothermal power plant, the largest such facility in Iceland. Hellisheiði is situated in a landscape of raw, elemental beauty where mats of chartreuse moss creep along cragged lava fields and sulfurous fumes rise from cracks in the earth. Contrasting with this primordial ambience is a series of ultramodern buildings, including a welcome center and museum with glass walls and a metal triangle angled off the roof like a giant compass needle. Nearby, the Swiss startup Climeworks has erected eight steel boxes, each about the size of a shipping container, stacked in pairs and arranged to form a square bracket. Each box houses what are effectively a dozen atmospheric vacuum cleaners continually sifting the sky.

On a misty September morning, I joined around two hundred scientists, investors, politicians, and journalists at Hellisheiði to witness the inauguration of Climeworks's Icelandic operations. Up close, we could see that large fans were embedded on one side of each steel box.

These fans, we learned, channeled air toward filters that captured carbon dioxide, allowing the decarbonized air to escape through vents. When a filter became saturated, heat was used to extract the CO_2 so that it could be collected and stored elsewhere. The process is called direct air capture because it absorbs carbon directly from the ambient air, as opposed to soaking up the fumes of a factory or a fossil fuel–burning power plant. This particular direct air capture facility, known as Orca, is powered almost entirely by geothermal energy.

The next stop on the tour was a nearby cluster of silver geodesic domes that looked like housing pods in a newly founded Martian colony. Employees of the Icelandic company Carbfix, which is collaborating with Climeworks, passed out neon yellow safety vests and white hard hats and invited us to look inside. Each dome sheltered a borehole and a system of large, interconnected metal pipes studded with rivets and handwheels. Sandra Snæbjörnsdóttir, a geologist with Carbfix, explained how pipes carried carbon dioxide captured by Orca to the domes, where it was mixed with water and injected thousands of feet below the surface into a layer of porous basalt. "Basalt is kind of like a sponge," Snæbjörnsdóttir said. "And it can hold a lot of CO_2."

When the carbon dioxide and water are discharged in the geothermal oven of Iceland's crust, they immediately begin reacting with certain elements in the basalt, forming chalky calcium carbonate that fills the bedrock's abundant pores and fissures within a matter of months. Molecule by molecule, air becomes stone and atmospheric carbon is once again locked in the planet's crust for tens of thousands to millions of years. Climeworks's Icelandic operations are the first manifestation of the company's true ambitions: to sell safe and effectively permanent carbon sequestration to individuals and corporations in order to offset their emissions.* So far, their roster of customers and investors includes Microsoft, Stripe, Shopify, Square, Audi, John Doerr, and Swiss Re.

* Scientists estimate that there are enough suitable and accessible geologic formations around the world to store many trillions of tons of CO_2, far more than humanity's entire historical output. Carbfix and independent researchers have determined that when geologic storage of carbon dioxide is properly managed with the appropriate safeguards, the risk of leakage or induced seismicity is minimal.

The morning after the launch of Orca, I sat down with Christoph Gebald and Jan Wurzbacher, the thirtysomething cofounders of Climeworks, in the swanky harborside startup space in Reykjavik where they were holding back-to-back conferences. They wore nearly matching outfits of slacks, dress shirts, and blue and gray sweaters. Both grew up in Germany and met in 2003 as engineering students at ETH Zurich, a research university focused on science and technology, where they bonded over their mutual difficulty understanding Swiss German and their shared ambition to head their own company. One of their professors introduced them to the work of Klaus Lackner, a distinguished physicist and the first person to formally propose direct air capture as a way to regulate atmospheric carbon. The idea captivated them.

"I was amazed by this prospect," Gebald told me in one conversation. "It sounded really relevant and very big. Today, smoking in an airplane feels wrong, right? I think in twenty years, people will feel the same about putting gasoline in cars or having coal-fired power plants."

As graduate students, Gebald and Wurzbacher developed a prototype consisting of little more than a bucket stuffed with filters coated in amines: compounds derived from ammonia that excel at snagging carbon dioxide from the passing air. Climeworks was incorporated as a university spinoff in 2009. At first, the cofounders had difficulty securing funding, especially after a 2011 assessment by the prestigious American Physical Society concluded that direct air capture was not economically viable. A windfall investment from a Swiss foundation helped them build a much more powerful device the size of a refrigerator, which became the basis for the world's first commercial direct air capture plant in Hinwil, Switzerland. The 2017 launch of that facility earned Climeworks new levels of fame and tens of millions of dollars of additional funding.

The following year, the IPCC released a report stating that in order to prevent global temperature from increasing more than 1.5°C (2.7°F) over the preindustrial baseline—the goal that nearly two hundred countries agreed to pursue during the 2015 Paris Climate Accords—the world would need to do more than drastically reduce the flow of carbon to the atmosphere. Humanity would also have to re-bury some of the carbon it had unearthed. The report mentioned several possible strategies to accomplish this, including direct air capture; reforestation; a technique known as bioenergy with carbon capture and storage (BECCS), which involves growing and burning plants to generate energy while simultaneously capturing any carbon emitted by their combustion and sealing it in the earth; and enhanced weathering, a process in which crushed basalt and other silicate rocks are sprinkled over land and sea to absorb CO_2.

Despite the IPCC's assessment, direct air capture and other forms of greenhouse gas removal, or negative emissions as they're also known, remain some of the most contentious ideas in climate science and policy. In order for the direct air capture community to meet its aspiration of removing several billion tons of carbon dioxide annually by 2050, it will need thousands of plants across the globe operating at a scale and efficiency orders of magnitude greater than anything demonstrated thus far. By some estimates, that would entail an entirely new industry capable of handling a volume of carbon comparable to or greater than that managed by the world's current oil infrastructure.

As of late 2023, Climeworks is one of only two companies operating a commercial direct air capture and storage facility, the other being Heirloom in California. When they finish constructing a new, much larger plant at Hellisheiði to supplement Orca, the site's total capacity will increase to forty thousand tons of CO_2 each year, which still amounts to only thirty-five seconds' worth of annual global emissions. Several other companies—including Carbon Engineering and Global Thermostat—have built pilot plants and, at the time of writing, are constructing commercial facilities, each of which will reportedly capture between two thousand and one million tons of CO_2 every year.

In its current form, direct air capture technology is expensive, energy-intensive, and resource-hungry.* Although a direct air capture

* Orca alone required between $10 and $15 million to build. Climeworks currently estimates a cost of $800 to capture one ton of carbon. Climate scientist Zeke Hausfather has calculated that, even if the price falls to $100 per ton, as hoped, it would cost around $22 trillion for direct air capture to decrease global temperature by one-tenth of one degree Celsius.

plant does not require arable land and takes up far less space than, say, a forest—while also being less susceptible to wildfires and extreme weather—it still demands large amounts of cement and steel, the production of which are environmentally detrimental. Carbon management expert Jennifer Wilcox has estimated that a typical direct air capture plant will need the equivalent of the energy supplied by 300 to 500 megawatts over the course of a year in order to capture a million tons of CO_2, enough energy to power tens to hundreds of thousands of homes in the same year. Beyond concerns about scale and cost, some scientists and activists condemn direct air capture as a moral hazard: a technofantasy that allows large corporations to continue business as usual and distracts from the primary challenge of reforming the world's energy supply. Climeworks has not yet partnered with any fossil fuel companies, but other direct air capture startups have.

Proponents counter that in addition to drawing down past emissions, direct air capture is also one of the best ways to compensate for continued emissions from long-distance shipping, air travel, and the mass production of steel, cement, and nitrogen fertilizer, all of which will remain dependent on fossil fuels for the foreseeable future because there are currently no alternative energy sources powerful enough to sustain them. Climate scientist Zeke Hausfather favors an approximate balance of reducing current emissions by more than 90 percent and removing the less than 10 percent that remains with a diverse complement of technological and nature-based approaches. Negative emissions technologies could also enable a form of climate reparations. Poor countries in the Global South, which have contributed the least to climate change, will suffer its worst consequences. Some economists and sociologists have argued that the Global North, which has contributed the majority of greenhouse gases in the atmosphere today, should assume the expense and responsibility of removing and sequestering historic emissions in order to mitigate extreme weather in the low-income countries least able to adapt to it and grant them more time to complete their transitions to clean energy.

The cofounders of Climeworks concede that direct air capture is "not the silver bullet, not at all," as Gebald told me. "It is a part of a portfolio," he continued. "You also need to plant trees, you need to have CO_2 bound in your soils, you need to do enhanced weathering." But they think that their critics undervalue the significance and potential of the technology they are developing. "I understand that looking at the numbers today might be a bit frustrating because four thousand compared to forty gigatons is a match in an eternity of darkness," Gebald said. "But there's also this tendency to completely underestimate the collective power of technology." "Things can happen much, much faster than we think," Wurzbacher added. "That's why we are a fan of tech solutions, because they are in particular prone to exponential development. You just need to kick it off."

As a concept, direct air capture with geologic storage is seductive. Even if it cannot undo all the ecological repercussions of past emissions, it is, to an extent, the literal reversal of anthropogenic climate change, using technology to pull humanity's excreted carbon out of the atmosphere and turn it back into solid earth. It is an ostensible offer of redemption: the opportunity for the world's wealthiest nations to reach into the sky, revoke some of their sins, and put them to rest. In practice, however, at least in the near future, filtering carbon dioxide out of the atmosphere is unlikely to be more than an accessory, albeit a potentially important one, to the central challenge of the current planetary crisis. Preserving a version of Earth that can still support us and the multitudes of nonhuman life all around us—a new chorus in Earth's ever-evolving song—requires nothing less than rapidly disassembling and rebuilding the energy infrastructure that powers modern human civilization and transforming the relationship between our species and the planet as a whole.

MOST FORMS OF RENEWABLE energy are as old as Earth itself.* Humans have been using them since long before recorded history. We've been bathing in hot springs and cooking over fires since the Stone Age. We

* The five main types of renewable energy are solar, wind, water or hydropower, geothermal, and bioenergy or biomass, which is typically obtained by burning recently living plants.

have harnessed wind and water to sail ships and grind grain for millennia. Yet renewable energy technology is arguably still in its infancy, long stunted by the supremacy of the fossil fuel industry—until recently.

As a wind turbine engineer, Yi Guo is one of the many thousands of people around the world currently reforming the way modern human civilization harnesses, generates, stores, and transports energy. On its own, this energy revolution cannot resolve the current planetary emergency, which encompasses many intersecting crises, including anthropogenic climate change, widespread destruction of vital ecosystems, a staggering loss of species, unprecedented levels of pollution, and egregious socioeconomic inequality. Reforming the world's energy infrastructure is one of the most important and urgent undertakings facing humanity, however, because it is essential to halting global warming and preventing its worst repercussions as well as a prerequisite for the possibility of one day returning to an atmospheric CO_2 level and average temperature characteristic of the preindustrial baseline.

Managing the climate crisis depends on three monumental tasks: drastically reducing the flow of greenhouse gases to the atmosphere, removing and sequestering excess atmospheric carbon, and adapting to climatic shifts that cannot be avoided. First and foremost: wealthy nations need to rapidly replace fossil fuels with a combination of renewable energy and nuclear power;* conserve energy and improve energy efficiency wherever feasible; and electrify homes, businesses, and transport, while also helping countries with less wealth and resources fortify themselves against the inescapable consequences of climate change and eventually complete their own energy transitions. Established cities must become denser, greener, and more navigable for pe-

* Although nuclear power is not usually classified as a renewable, many scientists and environmentalists argue that it is clean, sustainable, and essential to a low-carbon energy infrastructure. Nuclear power plants pose some risks to human health and the environment, but they are by far one of the safest forms of energy production. Through the combined effects of greenhouse gas emissions, air pollution, and accidents during extraction, transport, and maintenance, fossil fuels kill thousands of times more people per unit of energy than nuclear power.

destrians, cyclists, trains, and buses, while new urban centers should be built this way from the start. Homes and buildings require better insulation and more efficient heating and cooling systems. Preventing leaks of refrigerants, which are some of the most powerful greenhouse gases in existence, and ultimately replacing them with less harmful alternatives are critical steps. Humanity must commit to more sustainable agriculture, plant-centric diets, reduced use of fertilizer, and less food waste. The ecosystems on which our species and so many others rely—rainforests, temperate forests, and kelp forests; grasslands, savannas, and chaparral; peatlands, wetlands, mangroves, and coral reefs, to name a few—must be restored and protected. We need to use both technology and ecology to capture and permanently sequester several billion tons of carbon every year. And we should adapt to the inevitable through a multitude of strategies, such as coordinated migration of people and assisted migration of plants and animals, early warning systems and shelters for extreme weather, flood protection, prescribed burns to limit wildfires, enhanced water security, climate change–resilient crops, and more green roofs, trees, and gardens in urban spaces for improved shade, water retention, and carbon storage.

Scientists have understood the basic physics underlying the greenhouse effect for nearly two centuries. In 1912, *Popular Mechanics* magazine explained in layman's terms how burning coal trapped heat in the atmosphere, warning that "the effect may be considerable in a few centuries." Since at least the 1970s, fossil fuel companies have known that their products were heating the planet with potentially devastating repercussions, a conclusion based in part on research they had funded. Instead of heeding the mounting evidence, they deliberately concealed it and dedicated huge sums of money to undermining scientific consensus on climate change, confusing and misinforming the general public, and swaying politicians in order to maximize their profits. This legacy of propaganda and corruption is one of the many reasons that the United States and many other wealthy industrialized nations have failed to respond to the climate crisis with anything close to the urgency it requires, even in the face of increasingly robust and impassioned activism.

At this point, the likelihood of preventing the planet from heating 1.5°C (2.7°F) over the preindustrial baseline is vanishingly small. Achieving that goal would require halving global CO_2 emissions by 2030 and reaching net zero emissions not long after. But 1.5°C is not a magic boundary between security and calamity. It is a compromise that emerged from more than two decades of negotiation between the world's major political powers and its most vulnerable nations. The plain truth is that every fraction of a degree matters. Every bit of additional warming intensifies extreme weather, endangers human and nonhuman lives, and exacerbates an already tumultuous future. Every bit of prevented warming saves lives, spares untold suffering, and preserves a more livable world. Consider the difference between an increase in global temperature of 1.5°C and 2°C. At 1.5 degrees, all the sea ice in the Arctic will melt once every hundred years. At 2 degrees, it will happen once every ten years. At 1.5 degrees of warming, coral reefs will suffer a decline of 70 to 90 percent. At 2 degrees, it will be more than 99 percent. Compared with 1.5°C, 2°C means double or triple the rate of species loss, 2.6 times as many people exposed to extreme heat, and around twice as many people threatened with water scarcity.

Although humanity as a whole has not done nearly enough to manage the climate crisis, it would be unequivocally false to claim that there has been no meaningful progress, let alone that the world is doomed. Not only has there been measurable progress, but there are also good reasons to expect that it will accelerate. Some climate experts and activists are more optimistic today than they have been in a long time.

Before the 2015 Paris Accords, scientists predicted that by 2100 Earth's average temperature would increase between 4 and 5°C (7.2 to 9°F) over the preindustrial baseline—a calamity of unfathomable proportions. Since then, a lot has changed. The latest IPCC report estimates that an expansion of climate policies around the world has prevented several billion tons of CO_2 emissions each year. Current policies will result in around 3°C (5.4°F) of warming by the end of the century. That's still far too much, but it is inarguably progress. If the world's nations keep their current (nonbinding) promises to reduce emissions even further, they will restrict warming to between 2 and 2.4°C by 2100. If every country additionally meets their goals to reach net zero emissions by 2050, end-of-century warming should not exceed 2°C (3.6°F).

At the time of writing, global CO_2 emissions are still at historic levels in absolute terms, but their growth has slowed markedly over the past decade; some experts argue that they may soon peak. During the 2021 United Nations Climate Change Conference, also known as COP26, more than 130 countries, including Brazil, Canada, China, Indonesia, Russia, and the United States, pledged to "halt and reverse forest loss and land degradation by 2030," and more than one hundred countries agreed to cut methane emissions 30 percent by the same year. The conference culminated in the Glasgow Climate Pact, the first United Nations agreement in history to explicitly address the need to quit fossil fuels, calling upon the 194 parties to accelerate efforts toward the "phasedown of unabated coal power and phase-out of inefficient fossil fuel subsidies," while providing "targeted support to the poorest and most vulnerable."

In the past few decades, renewable energy technologies, in particular solar, wind, and batteries, have become cheaper much faster, and proliferated much sooner, than even the most sanguine experts predicted. In many if not most parts of the world, renewable energy is now more affordable than fossil fuels.* Globally, about 11 percent of primary energy and nearly 30 percent of electricity now come from renewables. At least sixty-five countries use renewables for more than half of the electricity they generate.†

* The pace of this change is staggering. The price per watt of energy generated by solar photovoltaics plunged by 99.6 percent between 1976 and 2019, from more than $106 to 38 cents. Similarly, the price of lithium ion batteries has declined by 97 percent over the past three decades. Since 2010, the average cost of wind energy has dropped between 55 and 70 percent.

† These pronounced shifts are partly due to advances in the underlying technologies, which have in turn been facilitated by government subsidies and both public and private research and development. The inherent intermittency of certain renewable energy sources is well known and is rapidly being solved with precise forecasts of generation patterns, enhanced storage, complementary combinations of different renewables, and better management of demand and supply through more thorough integration of energy systems.

As of 2019, clean energy jobs in the United States outnumbered jobs in the fossil fuel industry three to one. The U.S. Bureau of Labor Statistics predicts that wind turbine services will be the second-fastest-growing job in the country through 2030.* And as environmentalist Bill McKibben has noted, about $40 trillion worth of endowments, portfolios, and pension funds have now committed to full or partial abstinence from coal, gas, and oil stocks, a sum approximately equivalent to the gross domestic products of the United States and China combined.

On August 16, 2022, U.S. President Joe Biden signed the Inflation Reduction Act into law, dedicating $369 billion to building energy security and mitigating climate change—the largest such investment thus far. The law will subsidize electric cars, solar panels, and energy-efficient appliances; fund soil enrichment and sustainable agriculture; and offer tax credits for carbon capture and storage, among other provisions. By 2030, the law should drive down U.S. greenhouse gas emissions to between 30 and 40 percent below their 2005 levels, according to several independent research groups.

Environmental scientist Jonathan Foley has been studying Earth's climate since the 1980s and is now executive director of Project Drawdown, a nonprofit that strives to help the world reach the era-defining moment when "levels of greenhouse gases in the atmosphere stop climbing and start to steadily decline." "I am more optimistic now than I have ever been about climate," he says. "We're only doomed if we choose to be. We have solutions to stop climate change right in our hands. We know they can work. So what are we going to choose to do?"

THE CURRENT CLIMATE CRISIS is, fundamentally, the result of a major imbalance in the Earth system—one created entirely by our species.

* Unlike their predecessors, the newest generations of wind farms are designed and situated to minimize the risk of harming birds, bats, and other wildlife. Wind turbines kill an estimated 140,000 to 680,000 birds annually in the United States, but that number is dwarfed by the several billion birds killed every year by domestic cats and collisions with buildings, along with hundreds of millions of additional deaths due to cars and power lines. The Audubon Society strongly supports the development of properly managed wind energy.

Our planet tends toward radiative equilibrium, a state in which the energy it receives from the sun is equal to the energy it radiates back to space and global temperature remains relatively stable. By trapping heat that would otherwise escape, greenhouse gases knock Earth out of radiative balance, forcing its temperature to rise.

Like animism, balance is one of humanity's oldest and most universal concepts. Throughout history, people in many cultures have believed that the world is defined by, and dependent on, a balance of distinct and often opposing forces: light and dark, life and death, order and chaos. Such beliefs manifested themselves in the early development of Western science, especially in natural history. The ancient Greek historian Herodotus wrote that, through divine providence, predators were innately less prolific than their prey, which prevented them from hunting the latter to extinction. To support this assertion, he fabricated the example of lion cubs shredding their mothers' wombs with their claws so that they could never give birth again. Similarly, in 1714, English clergyman and natural theologian William Derham proclaimed, "The Balance of the Animal World is, throughout all Ages, kept even, and by a curious Harmony and just Proportion between the increase of all Animals, and the length of their Lives, the World is through all Ages well, but not over-stored." A few decades later, Carl Linnaeus, the Swedish scientist who formalized modern taxonomy, published an essay titled "Oeconomia Naturae," or the Economy of Nature, in which the term *economy* was a synonym for physiology, the study of how the different parts of a living system worked together to maintain overall well-being. The "divine wisdom," Linnaeus explained, had ensured that "all natural things should contribute and lend a helping hand toward preserving every species" and that "the death and destruction of one thing should always be subservient to the resolution of another."

Over time, the concept of balance in nature evolved, accommodating the dramatic and increasingly evident transitions in the epic history of life on Earth. Charles Darwin wrote of numerous checks on the short-term growth of living populations and regarded extinction as a gradual process balanced by the emergence of new, better adapted species. Herbert Spencer, a peer of Darwin who coined the phrase "survival of the fittest," proposed that every species experienced a rhythmical variation in numbers depending on the availability of food and the prevalence of environmental hazards. Amid these oscillations, he added, was an average number at which growth and decline were in equilibrium, meaning there was no net change one way or the other. In the early twentieth century, American ecologist Frederic Clements argued that, much like an individual organism, forests and other plant communities underwent a developmental process called succession, progressing from a juvenile stage to a mature "climax state" in which they were optimally adapted to the local environment. If anything disturbed that balance, the community attempted to restore it. Several decades later, Eugene Odum, often considered one of the founders of modern ecology, wrote that all living entities, from individual cells to entire ecosystems, had the capacity to maintain homeostasis.

By the latter part of the twentieth century, however, numerous ecologists had either challenged or outright rejected the notion of balance in nature, especially as it concerned the rise and fall of populations. Extensive research had established that predators and prey did not necessarily keep each other in check, that changes in species diversity and population growth were often wildly unpredictable, and that equilibrium or climax states, even if they could be mathematically modeled, were difficult to definitively identify in real-world ecosystems. Despite these findings, the idea that nature tends toward balance became thoroughly integrated in public consciousness, even as distaste for the concept intensified in academia. In a 2009 book, Wheaton College biology professor John Kricher called the balance of nature an "enduring myth" and ecology's "most burdensome philosophical baggage." In a historical review published in 2014, Daniel Simberloff, a professor of environmental science at the University of Tennessee, Knoxville, concluded that among professional ecologists, "the notion of a balance of nature has become passé, and the term is widely recognized as a panchreston—a term that means so many different things to different people that it is useless as a theoretical framework or explanatory device."

This umbrage is somewhat understandable. The classical idea of nature as a fixed arrangement, essentially perfect and unchanging, clearly does not reflect reality. Nor do the cartoonish depictions of ecological harmony sometimes encountered in popular culture, all sunshine and birdsong without a hint of conflict. Yet the complete repudiation of balance, much like the disparagement of the Gaia hypothesis, obscures important truths about the world. Our living planet abounds with examples of what can reasonably be called balance.

Predators and prey may not always achieve textbook equilibrium, but each continually evolves in response to the other, countering new hunting techniques with more effective defenses. Forests, grasslands, and coral reefs do not steadily advance toward some optimal climax state, but they can persist for tens of millions of years, retaining their essential qualities even as their compositions change. Some scientists have argued, for example, that the Amazon rainforest has endured for at least 55 million years and "should not be viewed as a geologically ephemeral feature of South America, but rather as a constant feature of the global Cenozoic biosphere." Patterns of speciation may not be wholly predictable, but given enough time and opportunity, species within a given ecosystem tend to fill a diverse complement of niches. Over great spans of time, anatomical features, ecological relationships, and even entire biomes evolve, disappear, and reemerge, either in a somewhat altered or uncannily similar form. And after every one of the five mass extinctions in its 4.5-billion-year history, each of which eradicated the majority of species in existence at the time, the planet not only rebounded but eventually prospered.

When most people speak of the balance of nature, I doubt that they mean a strict equilibrium or an unlimited capacity for recovery, as some scientists have implied. Rather, the "balance of nature" is typically shorthand for, as Rachel Carson phrased it, "a complex, precise, and highly integrated system of relationships between living things" that is "fluid, ever shifting, in a constant state of adjustment." Although this system is susceptible to disruption, it can also be restored or rearranged. "Balance" is meant to evoke this simultaneous intricacy, vulnerability, and resilience. Some scholars have argued that this characterization is self-contradictory, but complex living systems demonstrate precisely this multidimensionality.

The living entity we call Earth is in effect a highly complex balancing act sustained by the reciprocal evolution of organisms and their environments. Any living planet requires its animate and inanimate components to maintain certain relationships, rhythms, and cycles—a planetary physiology, so to speak. If Earth's atmosphere were in a state of perfect chemical equilibrium—like the atmospheres of Mars and Venus—it would not contain any free oxygen. Life pushed the atmosphere into a state of chemical disequilibrium that ultimately made the planet more habitable. In order to maintain that level of habitability, however, certain thresholds must not be crossed. Without sufficient oxygen in the sea and air, large complex life cannot exist. Too much, and the whole world erupts in flames. Not enough carbon dioxide in the atmosphere, and the planet freezes from pole to pole. An excess, and Earth becomes a hellish swampworld. The especially stable and clement version of Earth that our species and so many others have enjoyed these past twelve thousand years necessitates an even more specific set of environmental conditions.

For some time now, our planet has been moving toward a new equilibrium, a potential hothouse state, in which global temperature will be significantly higher and the consequently tempestuous climate will be devastating not only for human civilization but for scores of nonhuman species as well. If humanity continues to exhume and burn inordinate amounts of fossil fuels—thickening the planet's heat-trapping cloak and further imbalancing the Earth system—an appallingly inhospitable future is assured. If, instead, the nations most responsible for the climate crisis and most capable of resolving it finally act with the urgency it demands, they can still prevent global catastrophe. We may never be able to faithfully re-create the planetary rhythms and melodies of the past, but we don't need to. We can still perpetuate a rendition of Earth as we have known it—a variation on a theme.

BY THE TIME YI GUO completed her PhD in mechanical engineering, she was committed to a career in clean energy. She wanted to contribute to the rapidly evolving science that improved existing renewable energy technologies and conceived of new ones. She knew exactly where she wanted to work: the National Renewable Energy Laboratory in Colorado, NREL for short, which had a reputation as a world-class institute for wind energy research and development.

After a couple years of submitting applications, she landed a job as a postdoctoral researcher at NREL, where she eventually became a senior research scientist. Her studies focused on how best to extend the lifespan of wind turbines and enhance their overall reliability. Not long after her first day on the job, Guo ascended a wind turbine for the first time—an experience she remembers well. It was a three-hundred-foot-tall, three-megawatt turbine on NREL's Flatirons Campus, sufficient to power a small village. After safety training, she strapped on a harness and accompanied an experienced technician to the top.

"It was a thrilling experience," she says. "I was so excited. The wind can blow very suddenly and very strongly, but it's so beautiful up there. You feel so proud of what humans can achieve." The view was spectacular: below her, green fields from which sprang stately alabaster turbines; in the distance, the furrowed foothills of the Rockies, bristling with conifers; and all around her, a bright blue sky.

EPILOGUE

In South West England, on a stretch of the Jurassic Coast flanked by sea-sprayed pasture, there's a long and narrow gravel road running parallel to the beach. If you follow the road to its end, turn right, and walk slightly uphill, you will find a yellow brick cottage with a porthole embedded in the center of its front door. When I knocked on that door one fall morning, Sandy Lovelock answered. She was tall and thin with a well-tamed bob of snowy hair and a necklace of large canary-colored beads. James, who had recently turned one hundred, puttered along behind her in small shuffling steps, his kind brown eyes gleaming behind thick acrylic glasses.

Two decades earlier, while walking portions of the South West Coast Path—a 630-mile route from Somerset to Dorset—the Lovelocks had stayed at this same four-room cottage. When it finally went up for sale, they purchased it. "Now we're never without a view of the sea," James told me. We sat down to tea and biscuits in their living room, which was ornamented with curious souvenirs and gifts: an antique rocking horse from Paris, a nearly life-size wooden carving of a woman in a kimono, a card from Queen Elizabeth II. Through the living room window, I could see a small square lawn bordered by a low wall. In a bed at the bottom of the garden, Sandy had planted various palms and yuccas. Nestled among them was a stone statue of the Greek goddess Gaia.

On the wall, just to the right of where James was sitting, hung a large and colorful piece of art, halfway between a landscape painting and a surreal collage: mountains, valleys, and whipped mounds of clouds filled the background; lush rainforests and mangroves looked as though they might burst through the canvas; and a sumptuous swirl of ocean carried tropical fish, coral, and magnified plankton into the center of the frame. It was, I realized, a portrait of our living planet—of the overlapping ecosystems that sustain Earth as we know it.

The Lovelocks and I talked for several hours on many subjects: their lives and careers, the English countryside, recent books and movies, and childhood memories. Although physically frail, James was cheerful, articulate, and quick-witted. He thrived on humor, grinning broadly at even mildly amusing comments and hooting with laughter as he recounted favorite anecdotes. It was not long before we were discussing the genesis and evolution of the Gaia hypothesis. "Anything alive is capable of changing the planet," James said at one point. "That's what's fascinating about it." I asked him what it means to him to say that Earth itself is alive. It means, he said, that our planet is "similar to a living organism such as yourself, or a bacterium, or anything that maintains a regulated state of composition despite an environment which all the time is trying to undo it."

"And why do you think there was so much resistance to this idea from some sectors of the scientific community?" I asked. "Oh, the reason is quite simple and very human," James replied. "And it goes back to the Middle Ages or even earlier. 'We have this theory and it's like *this*. Don't talk to us about *that*.' Their careers, their livelihood depends on obeying those sorts of conditions." I mentioned that some prominent scientists had recently changed their minds about Gaia, including evolutionary biologist W. Ford Doolittle. "Yes, I'd heard that," James said. "They will all come around in time."

When James Lovelock passed away in the summer of 2022 at the age of a hundred and three, he left behind a long, illustrious, and complicated legacy. He was a man of prodigious intellect, numerous talents, and wide-ranging interests: a medical doctor, engineer, and author. For much of his career, he worked as an independent scientist rather than as a full-time employee of any one university or company, traveling the world and consulting for a variety of different organizations, including intelligence services and fossil fuel corporations.* He invented the electron capture detector, a small device of unprecedented sensitivity capable of detecting chemicals at concentrations as low as one part per trillion—a device that ultimately helped reveal the pervasiveness of pesticides and other environmental contaminants as well as the hole in the ozone layer. In the 1950s, while researching how to reanimate frozen rats and hamsters, he accidentally invented an early version of the microwave oven. And he published his first papers on the Gaia hypothesis several decades before Earth system science was an established field.

In subsequent talks and publications, Lovelock's concept of Gaia continually shifted. At times, he contradicted himself. In some of his writings, he explicitly stated that Gaia was a living entity, a vast being, or a superorganism; elsewhere, he said Gaia was only alive metaphorically, the same way a gene was selfish. He declared that humans "are no more qualified to be the stewards or developers of the Earth than are goats to be gardeners," yet simultaneously appointed himself a "planetary physician," prescribing his preferred treatments and urging everyone to participate in the necessary global healing. In some books, he made wildly hyperbolic predictions: "before this century is over, billions of us will die and the few breeding pairs of people that survive will be in the arctic region where the climate remains tolerable," he wrote in 2006 in *The Revenge of Gaia*. He later recanted those claims, calling them "alarmist." In *Novacene,* the last book he published before his death, Lovelock argued that Earth's future belongs to artificially intelligent cyborgs who will inevitably transcend human cognition. "If I am right about the Gaia hypothesis and the Earth is indeed a self-regulating system, then the continued survival of our species will depend on the acceptance of Gaia by the cyborgs," he wrote. "In their own interests, they will be obliged to join us in the project to keep the

* Historian Leah Aronowsky contends that Lovelock's consultancy for oil and gas company Shell influenced his early thinking on Gaia. For a discussion of the evidence, see the supplementary notes at the end of the book.

planet cool. They will also realize that the available mechanism for achieving this is organic life. This is why I believe the idea of a war between humans and machines or simply the extermination of us by them is highly unlikely. . . . We will be to the cyborgs as pets and plants are to us."

When Lovelock conceived of Gaia in the 1960s, he was thinking not about cyborgs but about aliens. Following the success of the exquisitely sensitive electron capture detector, NASA hired Lovelock to devise methods of discovering extraterrestrial life on Mars. While working at NASA's Jet Propulsion Laboratory in Pasadena, California, Lovelock had many stimulating conversations with colleagues in diverse fields, including cosmologist Carl Sagan, astronomer Lou Kaplan, and philosopher Dian Hitchcock. Together, Lovelock and Hitchcock determined that the most efficient way to detect the presence of life on another planet would be to analyze the chemistry of its atmosphere. Wherever life evolved, they reasoned, it would inevitably transform the rock, water, and air of its home planet. An intelligent alien examining Earth from afar would immediately recognize the unusual abundance of oxygen and other reactive elements in the atmosphere as a sign of life resisting entropy and pushing the planet into a state of chemical disequilibrium. Without photosynthetic life, Earth's atmosphere would be dominated by relatively inert carbon dioxide and contain essentially no free oxygen, just like the atmospheres of Mars and Venus. Earth's nearest siblings may have been alive at one point, but based on their atmospheric chemistry, Lovelock was confident that they were now lifeless—an assertion that has thus far held true.

Around the same time that Lovelock was working for NASA, astronaut Bill Anders took one of the first clear, color images of our entire planet suspended in space. When astronauts see Earth this way for the first time, they often experience a psychological phenomenon known as the overview effect: they are suddenly overwhelmed by the simultaneous beauty and vulnerability of the planet and a new understanding of their place in the cosmos. Space historian Frank White has said that people who go to space "will see things that we know, but that we don't experience, which is that the Earth is one system. We're all part of that system, and there is a certain unity and coherence to it all."

We now have ample evidence of this unity. More than 3.5 billion years ago, fragments of young Earth assembled themselves into the first genes, proteins, and cells. Microbes, the earliest outgrowths of our living planet—the smallest and most ancient of all organisms—intermingled as they adapted to Earth's primordial environments, exchanging DNA, consuming one another, and eventually merging into larger and more complex life forms. Since then, the tree of life has continually branched into innumerable species adapted to diverse habitats distributed throughout the planet's solid, liquid, and gaseous layers. As they evolved, those creatures profoundly altered their environments, too, oxygenating the atmosphere, redefining ocean chemistry, and turning barren crust into fertile soil. The linear, branching evolution of species has always been embedded in the looping, reciprocal evolution of life and environment.

We need to peel Darwin's evolutionary tree of life off the page and reanimate it as a gloriously convoluted four-dimensional entity. Imagine a massive tree with a tangled web of roots and overlapping branches that sometimes fuse—a physical manifestation of life's ongoing evolution. Imagine, too, that as the tree grows, it continuously changes its surroundings, burrowing into rock with vigorous roots, enriching dirt with cast-off remains, remaking the air, seeding clouds, and summoning rain. In turn, these environmental transformations influence the tree's continued growth. Thus are the evolution of life and environment—of Earth and its creatures—bound together. Living networks that destroy their environments ultimately doom themselves, whereas those that maintain and improve their surroundings are more likely to endure.

In this way, eons of coevolution of Earth and life seem to have favored ecological relationships that enhanced the planet's habitability, endowing it with features resembling those of a living organism, including a capacity for self-regulation, albeit constrained. Even today, some scientists, especially in geology and related fields, continue to describe life as a thin layer of goo covering the planet. But that characterization belies life's true scale and power. Life dramatically expands the surface area of the planet capable of absorbing energy, exchanging gases, and performing complex chemical transformations. Earth system scientist Tyler Volk has calculated that the surface area of all the sun-sponging photosynthetic plankton in the ocean is six times greater than the planet's inanimate surface area. All the plant roots on Earth, finely furred with tiny absorptive hairs, comprise a surface area thirty-five times greater than the planet would have without life. Microbes are collectively equivalent to two hundred Earth areas. And if there were a layer of fertile soil three feet thick spread across the continents, all the tiny particles within it would have a combined surface area more than a hundred thousand times that of the bare planet. There's simply no comparison. Life gives our planet an anatomy and physiology—breath, pulse, and metabolism. Without the transformations wrought by life over billions of years, Earth would be utterly unrecognizable. Life does not merely exist *on* Earth—life *is* Earth. We have as much reason to regard our planet as a living entity as we do ourselves—a truth no longer substantiated by intuition alone, nor by one man's vision, but by a preponderance of scientific evidence.

Like any long-lived creature, Earth's life has been tumultuous. Throughout Earth's history, catastrophes of incomprehensible scale have repeatedly punctuated periods of relative stability. Some of those crises were primarily caused by geological or cosmological events, such as volcanic eruptions and asteroid strikes; others were mostly the result of evolution's many experiments. In each case, despite the extinction of most species that existed at the time, our living planet gradually recovered over many thousands to millions of years, often becoming a radically different world. A living planet, it seems—especially one that has evolved a high level of ecological complexity—has the potential for extraordinary resilience over geological timescales. But the self-stabilizing processes underlying this resilience operate much too slowly, and involve far too much upheaval, to spare such transient entities as human societies.

We are currently living through a planetary emergency—one of our own making. By burning immense stores of fossil fuels, destroying forests and grasslands, and polluting the air, sea, and land, among many other forms of disruption, we have pushed Earth into yet another crisis. Although it may not be the most severe disaster in the planet's history, some aspects of it are likely unprecedented, including the speed with which we have flooded the atmosphere with such a large volume of carbon. Without the necessary interventions, the planet will become inhospitable not only to humans but also to countless forms of complex life.

While writing this book, I frequently encountered three opposing perspectives on the fate of humanity and our planet. One camp, which we might call the fatalists, thinks that wherever intelligent, self-aware, and technologically skilled life evolves, it is doomed to extinguish itself and destroy its home planet. Why is it, they ask, that we have never encountered intelligent alien life despite inhabiting such a vast universe in which we are surely far from alone? Because, they answer, spacefaring species never last long enough to meet each other. In contrast, the fantasists regard the living planet as a benevolent being, continually moving toward a state of zen perfection. In their view, there is little need for alarm, because the planet will ultimately take care of itself, which may or may not entail shaking loose a certain pesky simian species. Members of the third group, the futurists, generally foresee a bleak future for Earth but reject both apathy and defeatism, determined to find or create a new home for humanity on a different planet.

I would like to suggest an alternative perspective informed by Earth system science. Life is not a wholly benevolent force purposefully working toward the greater good. There is no optimal state for the planet. Yet it is true that over great spans of time, Earth and its creatures have tended to coevolve relationships that promote their mutual persistence and imbue the planet with remarkable tenacity. Let's be clear: we are not in danger of killing the creature we call Earth. Even if we unearthed and burned all the fossil fuels in existence—even if we induced a hellish hothouse state that extinguished our species and the majority of complex life—microbes and other resilient forms of life would persevere and the planet as a whole would eventually recover.

What we are in the process of destroying is the world as we have known it: the particular version of Earth in which our species and so many others evolved—a rendition of Earth that, compared to many of its previous states, is a genuine Eden. If unchecked, the horrifying transformation we have set in motion will ravage ecosystems across the globe and ruin billions of lives. The interventions required to prevent the worst outcomes of this crisis and maintain a habitable planet are known and achievable. If the nations most responsible for the current crisis, and most capable of managing it, fail to uphold their responsibilities, they will sacrifice much more than the Earth we've known—they will also preclude the possibility of a better world for humanity. To ignore the necessary changes here on Earth in favor of terraforming other planets in any meaningful time frame is unforgivable folly. We are nowhere near the level of ecological understanding and technological sophistication required to turn an inanimate and airless rock into a new Earth, but we are without question capable of preserving the one living planet we already have—and the only one we've ever found.

In his last book, Lovelock argued that the "staggeringly improbable chain of events required to produce intelligent life" has occurred once and only once in the known cosmos—that humanity's existence is "a freakish one-off." Considering the mind-boggling size and age of the observable universe, however, it seems highly unlikely that our species is a complete anomaly. With an estimated two trillion galaxies in the cosmos, and perhaps tens of billions of habitable planets in our galaxy alone, there is undoubtedly life beyond Earth. Aliens are real. The vast majority of life in the universe may be small and simple, analogous to Earth's single-celled microbes. If our planet's history is anything to judge by, complex life needs a lot of time and opportunity to emerge. The sheer number of planets in the habitable zones of the stars they orbit provides that opportunity. And at 4.54 billion years and counting, Earth is not a particularly old planet—only about one third the age of the universe. There may well be planets on which life has been evolving somewhere between 5 and 13 billion years.

If there are other highly intelligent, self-aware, spacefaring life forms out there, perhaps the main reason we have not yet encountered them is that our already vast universe is not only expanding but expanding faster by the moment. Kim Stanley Robinson's novel *Aurora* tells the story of a twenty-sixth-century multigenerational starship's disastrous attempt to establish a human colony on a habitable moon twelve light-years from Earth. At one point, Aram, one of the survivors who makes it back to Earth, gives a speech: "The distances between here and any truly habitable planets are too great," he says. "And the differences between other planets and Earth are too great. Other planets are either alive or dead. Living planets are alive with their own indigenous life, and dead planets can't be terraformed quickly enough for the colonizing population to survive the time in enclosure. . . . That's why you aren't hearing from anyone out there. That's why the great silence persists. There are many other living intelligences out there, no doubt, but they can't leave their home planets any more than we can, because life is a planetary expression, and can only survive on its home planet."

I suspect that, should our species survive long enough—not just for several more centuries or millennia but into the inconceivably distant future—we will eventually learn to transform and inhabit other planetary bodies. If we do, Earth may become one of very few planets to successfully reproduce. But that possible future is entirely contingent on the decisions humanity makes right now about our current home. Earth has shown us the power of community, diversity, and reciprocity. Among all existing creatures, we alone have the opportunity to consciously emulate our living planet and knowingly perpetuate its sublime composition. We are neither the cancer of Earth nor its cure. We are its progeny, its poetry, and its mirror.

SEVERAL YEARS AGO, OREGON experienced one of its most severe winter storms in the past four decades. I woke up to a world recast in alabaster and glass. In the early morning light, treetops glistened like chandeliers of frost. The streets were sheened with white and silver. Every branch and twig was cocooned in ice.

The scene's beauty was rivaled only by its silence. Nearly every living thing appeared to be paralyzed. I was particularly concerned about our garden, which was not yet six months old. Many of our smallest and most tender plants were hidden beneath thick layers of ice and snow. The pond's once-bubbling surface was gray-blue slush. The bordering rushes looked as brittle as blown glass. Branches on our neighbors' trees sagged beneath the weight of their frigid casings, nearly touching the ground. Walking through my neighborhood felt like wandering a town bound by some chilling enchantment, populated only by the icy likenesses of the once living.

Looking back now, I see things differently. Life was everywhere that winter morning. Part of the reason the trees and plants were so crystalline was that they harbored ice-making bacteria. The snow teemed with microscopic creatures that had evolved to survive the journey from ground to cloud and back again—creatures that not only endured the weather but also changed it. Beneath my feet, networks of root and fungi, and the multitudes of microanimals they attracted, were still respiring and growing. Some of the bulbs and tubers I had planted were already pulsing with self-generated heat, preparing to melt the snow and spear the soil. The many coniferous trees in my neighborhood continued to pull liquid water from deep underground, collect sunlight through frost-sheathed needles, and pour oxygen into the atmosphere. Every element of winter was alive: the ice, the ground, the very air. The world was singing, even if I couldn't hear it.

I exhaled deeply, watching the ghost of my breath materialize: a shape-shifting cloud of water, gas, and cell. It billowed briefly, then dissolved. Borrowed elements returning to the source. Another note exchanged in my personal duet with the planet. I breathed out, and Earth breathed in.

ACKNOWLEDGMENTS

BEHIND EVERY BOOK, THERE IS AN ECOSYSTEM. THE PERSONAL curiosity quest that culminated in this book began more than a decade ago. In that time, I have been extremely fortunate to benefit from the assistance of numerous individuals and communities. Much of the writing I do depends on the hard work, knowledge, and generosity of scientists and other experts. I am deeply grateful to everyone I interviewed while writing this book, especially those who allowed me to join them in their homes and workplaces and on field expeditions. To the individuals I was not able to quote directly, please accept my sincere apologies and know that our interactions were nonetheless invaluable; every writer could fill a royal library with the material they were forced to leave off the page. Thank you to all the experts who responded graciously and insightfully to out-of-the-blue emails from someone they had never met. Thank you, too, to the many scientists whose published studies and books formed the foundation of my research.

As someone who frequently writes about science for a general readership, a balance of precision and clarity is of the utmost importance to me. My immense gratitude to the four professional fact-checkers whose meticulous work greatly improved the accuracy of this book: Jane Ackermann, Michelle Ciarrocca, Tina Knezevic, and Steven Stern. Thank you, also, to the many experts who fielded fact-checking queries and reviewed excerpts of the chapters, including Shady Giada

Anayati, Gaëtan Borgonie, Priyadarshi Chowdhury, Curtis Deutsch, Erle Ellis, Paul Falkowski, Gavin Foster, Geoffrey Gadd, Nicolas Gruber, Robert Hazen, James Kasting, Jun Korenaga, Lee Kump, Helmut Lammer, Tim Lenton, John Luczaj, Jennifer Macalady, George McGhee, Massimo Pigliucci, Simon Poulton, Chris Reinhard, Gregory Retallack, Andy Ridgwell, Patrick Roberts, Felisa Smith, Gordon Southam, Steven Stanley, Alexis Templeton, Tyler Volk, Andrew Watson, Jennifer Wilcox, Bruce H. Wilkinson, Mark Williams, Jan Zalasiewicz, and Richard Zeebe.

My editor, Hilary Redmon, encouraged me to pursue this project from its most embryonic stages and was a patient, enthusiastic, and incisive collaborator throughout the long process. My agents, Larry Weissman and Sascha Alper, helped me navigate the world of book publishing with skill and grace and have remained some of this book's staunchest advocates. A few passages in this book were originally published in *The New York Times* and *The New York Times Magazine*. Thank you to my editors at those publications for their support and collaboration, in particular Willy Staley, Jake Silverstein, Bill Wasik, Jessica Lustig, and Jeannie Choi. The reporting for this book was supported in part by a Whiting Creative Nonfiction Grant and an MIT Knight Science Journalism Fellowship. My sincere thanks to my fellow fellows and grantees for their counsel and solidarity as well as to the esteemed directors of those programs and their colleagues, especially Daniel Reid, Courtney Hodell, Deborah Blum, and Ashley Smart. I am likewise indebted to the entire team at Random House, including Evan Camfield, Toby Ernst, Erica Gonzalez, Michael Hoak, Miriam Khanukaev, Greg Kubie, and Alison Rich, and all the book's international publishers.

In the course of writing this book, several literary and professional communities were an indispensable source of camaraderie, advice, inspiration, and gleeful procrastination. Thank you in particular to all the wonderful members of the Slackline and Books Club, aka Creature Club. Heaps of thanks to my friends and literary peers who agreed to be early readers of the manuscript and whose insights immeasurably improved the text: Rebecca Altman, Rebecca Boyle, Emily Elert, Rebecca Giggs, Ben Goldfarb, Mara Grunbaum, Holly Haworth, Brandon Keim, Robert Moor, and Sierra Crane Murdoch. I am additionally grateful to the many friends who helped me workshop various aspects of the book and whose feedback was key to several important decisions: Ariel Bleicher, Nadege Dubuisson, Michael Easter, Caroline Foley, Ian and Bekka Gillman, Kei Higaki, David and Leah Jobson, Olivia Koski, Reid Koster, Alex Liu, Dylan and Taylor McDowell, Ryan and Annie McMahon, Erin Mellon, Mike Orcutt, Katie Peek, Anna Rothschild, Nicole Sharpe, and Josh and Shawnee Tracy. Much thanks, as well, to Mike and Masha Freeman for their linguistic expertise and assistance with translation.

It was a pleasure to collaborate with the multitalented Matthew Twombly, whose facility with art and science turned a compilation of research notes into the beautifully illustrated timeline of Earth history featured in the book. Thank you, likewise, to all the individuals who permitted me to use their photos.

Researching this book involved numerous reporting trips, some of which required extensive international travel. In an effort to compensate for the resulting greenhouse gas emissions, I am donating a portion of the advance I received for this book to organizations that help mitigate and manage the current planetary crisis, including the Indigenous Environmental Network, the Coalition for Rainforest Nations, the Clean Air Task Force, and Carbon180.

Thank you to my parents and siblings, who have cheered me on in all my pursuits since childhood. My partner, Ryan, has witnessed the joys, lulls, and travails of my book-writing journey more closely than anyone. Without his constant encouragement, saintly patience, and thoughtful feedback, I would have given up long ago. I owe a great deal to our dog, Jack, not only for his steadfast companionship but also for his insistence on numerous daily walks, which have become the metronome of my life and an integral component of my physical and mental well-being. Finally, and most fundamentally, my eternal thanks to the extraordinary creature we call Earth. May we be worthy of the life we share with you.

AUTHOR'S NOTE

THE EVOLVING DEFINITION OF GAIA

Earth scientist James Kirchner has developed a taxonomy of the many different versions of the Gaia hypothesis published over the years. What he calls Influential Gaia, the mildest version, asserts that life has a substantial influence on many aspects of the planet, such as the average global temperature and the chemical composition of the atmosphere. Similarly, Coevolutionary Gaia emphasizes that life and the planet change each other through reciprocal evolution. Homeostatic Gaia and Geophysiological Gaia, stronger forms of the hypothesis, claim that life tends to stabilize the planet and that Earth either is an immense living being or can be compared to one. Teleological Gaia and Optimizing Gaia, the most extreme versions, hold that life purposefully alters the planet in order to maintain the optimal conditions for itself. References to Kirchner's papers are listed below.

Kirchner and other scientists have also questioned whether the Gaia hypothesis qualifies as a hypothesis at all. In science, a hypothesis is a prediction or tentative explanation that can be tested through controlled experiments. While that definition may apply to specific proposals within the larger body of thought and scholarship now known as Gaia theory, it is far too narrow to capture it as a whole. Gaia is closer to a conceptual framework than a strict hypothesis. "The truth is, despite its widespread moniker, Gaia is not really a hypothesis," writes astrobiologist David Grinspoon in *Earth in Human Hands*. "It's a perspective, an approach from within which to pursue the science of . . . a living planet, which is not the same as a planet with life on it—that's really the point, simple but profound. Because life is not a minor afterthought on an already functioning Earth, but an integral part of the planet's evolution and behavior."

Here are a few key examples of how Lovelock and Margulis defined and redefined Gaia throughout their careers:

> "The purpose of this letter is to suggest that life at an early stage of its evolution acquired the capacity to control the global environment to suit its needs and that this capacity has persisted and is still in active use. In this view the sum total of species is more than just a catalogue, 'The Biosphere,' and like other associations in biology is an entity with properties greater than the simple sum of its parts. Such a large creature, even if only hypothetical, with the powerful capacity to homeostat the planetary environment needs a name; I am indebted to Mr. William Golding for suggesting the use of the Greek personification of mother Earth, 'Gaia.' "—Lovelock, "Gaia as Seen Through the Atmosphere," 1972
>
> ". . . the presence of a biological cybernetic system able to homeostat the planet for an optimum physical and chemical state appropriate to its current biosphere . . ."—Lovelock, "Gaia as Seen Through the Atmosphere," 1972
>
> ". . . the hypothesis that the total ensemble of living organisms which constitute the biosphere can act as a single entity to regulate chemical composition, surface pH and possibly also climate. The notion of the biosphere as an active adaptive control system able to maintain the Earth in homeostasis we are calling the 'Gaia' hypothesis."—Lovelock and Margulis, "Atmospheric Homeostasis by and for the Biosphere: The Gaia Hypothesis," 1974
>
> ". . . the proposition that living matter, the air, the oceans, the land surface were parts of a giant system which was able to

control temperature, the composition of the air and sea, the pH of the soil and so on so as to be optimum for survival of the biosphere. The system seemed to exhibit the behaviour of a single organism, even a living creature."—Lovelock and Epton, "The Quest for Gaia," 1975

"But if Gaia does exist, then we may find ourselves and all other living things to be parts and partners of a vast being who in her entirety has the power to maintain our planet as a fit and comfortable habitat for life."—Lovelock, *Gaia: A New Look at Life on Earth,* 1979

". . . the hypothesis that the entire range of living matter on Earth, from whales to viruses, and from oaks to algae, could be regarded as constituting a single living entity, capable of manipulating the Earth's atmosphere to suit its overall needs and endowed with faculties and powers far beyond those of its constituent parts."—Lovelock, *Gaia: A New Look at Life on Earth,* 1979

". . . a complex entity involving Earth's biosphere, atmosphere, oceans, and soil; the totality constituting a feedback or cybernetic system which seeks an optimal physical and chemical environment for life on this planet. The maintenance of relatively constant conditions by active control may be conveniently described by the term 'homoeostasis.' "—Lovelock, *Gaia: A New Look at Life on Earth,* 1979

"As first proposed the Gaia hypothesis was, I accept, badly worded. . . . Unwisely we stated the hypothesis as 'Life or the biosphere modulates the environment to keep it comfortable' or 'The biosphere sustains the environment by and for itself.' We should have said, 'Living organisms and their material environment are tightly coupled. The coupled system is a superorganism, and as it evolves there emerges a new property, the ability to self-regulate climate and chemistry.' It took ten years and a mathematical model, 'Daisyworld,' to define Gaia, the superorganism."—Lovelock, *The Ages of Gaia,* 1988

"The nearest I can reach is to say that Gaia is an evolving

system, a system made up from all living things and their surface environment, the oceans, the atmosphere, and crustal rocks, the two parts tightly coupled and indivisible. It is an 'emergent domain'—a system that has emerged from the reciprocal evolution of organisms and their environment over the eons of life on Earth. In this system, the self-regulation of climate and chemical composition are entirely automatic. Self-regulation emerges as the system evolves. No foresight, planning or teleology . . . are involved."—Lovelock, *Healing Gaia: Practical Medicine for the Planet,* 1991

"The concept 'Gaia,' an old Greek name for Mother Earth, postulates the idea that the Earth is alive. The Gaia hypothesis, proposed by the English chemist James E. Lovelock, is that aspects of the atmospheric gases and surface rocks and water are regulated by the growth, death, metabolism, and other activities of living organisms."—Margulis, *Symbiotic Planet: A New Look at Evolution,* 1996

"The Gaia hypothesis is not, as many claim, that 'the Earth is a single organism.' Yet the Earth, in the biological sense, has a body sustained by complex physiological processes. Life is a planetary-level phenomenon and Earth's surface has been alive for at least 3,000 million years."—Margulis, *Symbiotic Planet: A New Look at Evolution,* 1996

"As detailed in Jim's theory about the planetary system, Gaia is not an organism. Any organism must either eat or, by photosynthesis or chemosynthesis, produce its own food. . . . Gaia, the system, emerges from ten million or more connected living species that form its incessantly active body. . . . As they unwittingly obey the second law of thermodynamics, all beings seek energy and food sources. All produce useless heat and chemical waste. This is their biological imperative. . . . The sum of planetary life, Gaia, displays a physiology that we recognize as environmental regulation. Gaia itself is not an organism directly selected among many. It is an emergent property of interaction among organisms, the spherical planet on which they reside, and an energy source, the sun."—Margulis, *Symbiotic Planet: A New Look at Evolution*, 1996

"Gaia is the series of interacting ecosystems that compose a single huge ecosystem at the Earth's surface. Period."—Margulis, *Symbiotic Planet: A New Look at Evolution*, 1996

LOVELOCK'S CONSULTANCY FOR SHELL

In 1963, Lovelock decided to become an independent scientist. Instead of working full-time for a single institution, he started to support himself by consulting for various universities, companies, and organizations. One of those organizations was NASA's Jet Propulsion Laboratory in Pasadena, California. It was in 1965, on one of his visits to JPL, that Lovelock had an epiphany about life altering and regulating the planet—the seed that grew into the Gaia hypothesis. In 1966, Lovelock began an additional consultancy for Shell Research Limited, the research branch of oil and gas giant Royal Dutch Shell, which asked him to investigate the "possible global consequences of air pollution from such causes as the ever-increasing rate of combustion of fossil fuels."

Lovelock has written openly about his work for Shell in several publications, emphasizing that it was not a "possessive relationship" and that he maintained "freedom of thought." Given how much he valued intellectual independence and originality and his penchant for puckishness and provocation, that is not hard to believe. As Lovelock wrote in *Gaia,* "The link between my involvement in the problems of global air pollution and my previous work on life detection by atmospheric analysis was, of course, the idea that the atmosphere might be an extension of the biosphere." His initial report for Shell concluded, in part, that it was an "almost certain fact" that the climate was worsening and that the combustion of fossil fuels was the most likely cause. In follow-up studies sponsored by Shell, he investigated the possibility that marine algae altered the atmosphere by producing dimethyl sulfide. In 1975, Lovelock co-authored an early article on Gaia in *New Scientist* with Shell manager Sidney Epton.

Historian Leah Aronowsky has argued, based in part on original archival research, that Lovelock's consultancy for Shell had a greater influence on the development of the Gaia hypothesis than is typically acknowledged and that Lovelock's ideas facilitated certain forms of climate change denialism. "Gaia," Aronowsky writes, "is a story in which a theory about the Earth's climate was put into the world that promptly made a range of new knowledge claims possible—including claims about the self-regulating stability of the climate later harnessed to sow doubt about global warming. . . . Simply put, Gaia created the conditions for a denialism that derived its power by denying the uniqueness of humans' capacity to permanently alter the Earth." Aronowsky also clarifies that "it is not the case that Gaia was a direct outgrowth of the fossil fuel industry's concerted campaigns to produce uncertainty about the scientific consensus on global warming" and that "it remains unclear the extent to which, at this stage in his [early] thinking, Lovelock drew a connection between his ideas about the cosmic signal of life and his work for Shell."

What is indisputably true is that as Gaia became increasingly popular, many individuals and organizations—including the fossil fuel industry—seized it and molded it to their various and often conflicting purposes. "Two groups that immediately embraced Gaia were environmentalists and, paradoxically, industrialists," Kirchner wrote in 1989. "The former argued that harming any part of the planetary 'organism' could have far-reaching consequences, while the latter argued that Gaia's capacity for homeostasis made pollution control unnecessary." In this way, Gaia became a tool of climate change denialism. Lovelock's own perspective on humanity's relationship with the planet continually shifted throughout his life; he sometimes got the facts wrong, contradicted himself, dismissed or exaggerated environmental threats, and made prejudiced or insensitive statements that many found morally repugnant. But he also fully accepted the reality of anthropogenic climate change and ultimately advocated for the end of fossil fuels.

THE GLOBAL CARBON MARKET AND THE FATE OF RUNNING TIDE

I visited Running Tide's headquarters in late May 2021. In subsequent years, the company achieved several new milestones, such as securing a deal with Microsoft to be the tech giant's first open-ocean-based carbon removal supplier and reportedly sinking more than 25,000 tons of biomass off the coast of Iceland. In parallel, however, scientists and journalists continued to raise a multitude of concerns about the feasibility, cost, and ecological risks of various carbon removal ventures, including Running Tide. By late 2023, many of the major corporations that had initially supported such projects became much more cautious and selective in their investments. In June 2024, by which point the first edition of this book had been finalized and printed, Running Tide abruptly announced that it had started to shut down its global operations, citing an inability to secure sufficient financing to continue its efforts at a meaningful speed and scale.

Some experts in the carbon dioxide removal community think that the many challenges and unanswered questions surrounding Running Tide's particular methods contributed to their closure. The company's strategies had evolved since my visit, placing new emphasis on enhancing the ocean's alkalinity by introducing limestone and similar minerals, in addition to sinking algal biomass. Yet experts simultaneously acknowledge that such failures are inevitable and necessary to finding the most effective approaches to long-term carbon sequestration. "Starting a company of any kind is a hopeful act," Running Tide founder Marty Odlin wrote on LinkedIn in June 2024. "Starting a company with the explicit goal of restoring nature and healing the ocean requires hope for the future and humility in the face of the complexity and the scale of the problem. I still have hope." A couple months later, Odlin bought back many of Running Tide's assets—including its name, IP, equipment, and data—with the goal of continuing its core mission, likely in the form of a leaner, more cost-effective organization. When I last corresponded with him in October 2024, he was working with around a dozen former employees to figure out the next steps. "I think we all feel confident we can help the world if we keep going," he told me, "but just trying to match the mode to the need."

SELECTED SOURCES

THE FOLLOWING IS A SELECTION OF SOME OF THE MOST IMPORTANT sources I consulted while writing this book, organized by chapter. Because the introduction and epilogue draw on the same body of research, I have combined their references below.

INTRODUCTION AND EPILOGUE

Aronowsky, Leah. "Gas Guzzling Gaia, or: A Prehistory of Climate Change Denialism." *Critical Inquiry,* vol. 47, no. 2, 2021, pp. 306–27.

Brannen, Peter. *The Ends of the World: Volcanic Apocalypses, Lethal Oceans, and Our Quest to Understand Earth's Past Mass Extinctions*. Ecco, 2017.

Carson, Rachel. *Silent Spring: Fortieth Anniversary Edition*. Mariner Books, 2002. (First published in 1962.)

Clarke, Bruce. *Gaian Systems: Lynn Margulis, Neocybernetics, and the End of the Anthropocene*. University of Minnesota Press, 2020.

Dessler, Andrew. *Introduction to Modern Climate Change: Second Edition*. Cambridge University Press, 2016.

Doolittle, W. Ford. "Is the Earth an Organism?" *Aeon,* December 2020.

Flannery, Tim. *Here on Earth: A Natural History of the Planet*. Grove Press, 2010.

Frank, Adam. *Light of the Stars: Alien Worlds and the Fate of the Earth*. W. W. Norton, 2018.

Grinspoon, David. *Earth in Human Hands: Shaping Our Planet's Future*. Grand Central Publishing, 2016.

Hawken, Paul, editor. *Drawdown: The Most Comprehensive Plan Ever Proposed to Reverse Global Warming*. Penguin Books, 2017.

Kimmerer, Robin Wall. *Braiding Sweetgrass: Indigenous Wisdom, Scientific Knowledge, and the Teachings of Plants*. Milkweed Editions, 2015.

Kirchner, James W. "The Gaia Hypothesis: Can It Be Tested?" *Review of Geophysics,* vol. 27, no. 2, 1989, pp. 223–35.

Kirchner, James W. "The Gaia Hypotheses: Are They Testable? Are They Useful?" *Scientists on Gaia,* edited by Stephen H. Schneider, MIT Press, 1991, pp. 38–46.

Kirchner, James W. "Gaia Hypothesis: Fact, Theory, and Wishful Thinking." *Climatic Change,* vol. 52, 2002, pp. 391–408.

Latour, Bruno. *Facing Gaia: Eight Lectures on the New Climatic Regime.* Translated by Catherine Porter. Polity Press, 2017.

Lenton, Tim. *Earth System Science: A Very Short Introduction.* Oxford University Press, 2016.

Lenton, Tim, and Andrew Watson. *Revolutions That Made the Earth.* Oxford University Press, 2014.

Lovelock, James. "Gaia as Seen Through the Atmosphere." *Atmospheric Environment,* vol. 6, no. 8, 1972, pp. 579–80.

Lovelock, James. *Gaia: A New Look at Life on Earth.* Oxford University Press, 1979.

Lovelock, James. *The Ages of Gaia: A Biography of Our Living Earth.* Revised and expanded edition. W. W. Norton, 1995. (First published in 1988.)

Lovelock, James. *Healing Gaia: Practical Medicine for the Planet.* Harmony Books, 1991.

Lovelock, James. *The Revenge of Gaia: Earth's Climate in Crisis and the Fate of Humanity.* Basic Books, 2006.

Lovelock, James. *The Vanishing Face of Gaia.* Basic Books, 2009.

Lovelock, James, with Bryan J. Appleyard. *Novacene: The Coming Age of Hyperintelligence.* MIT Press, 2019.

Lovelock, James, and Sidney Epton. "The Quest for Gaia." *New Scientist,* 1975.

Lovelock, James, and Lynn Margulis. "Atmospheric Homeostasis by and for the Biosphere: The Gaia Hypothesis." *Tellus,* vol. 26, 1974, pp. 2–10.

Margulis, Lynn. *Symbiotic Planet: A New Look at Evolution.* Basic Books, 1998.

Morton, Oliver. *Eating the Sun: How Plants Power the Planet.* Harper Perennial, 2009.

Skinner, Brian J., and Barbara W. Murck. *The Blue Planet: An Introduction to Earth System Science: Third Edition.* Wiley, 2011.

Smith, Eric, and Harold J. Morowitz. *The Origin and Nature of Life on Earth: The Emergence of the Fourth Biosphere.* Cambridge University Press, 2016.

Stanley, Steven M., and John A. Luczaj. *Earth System History: Fourth Edition.* W. H. Freeman, 2015.

Volk, Tyler. *Gaia's Body: Toward a Physiology of Earth.* MIT Press, 2003.

Ward, Peter, and Joe Kirschvink. *A New History of Life: The Radical New Discoveries About the Origins and Evolution of Life on Earth.* Bloomsbury Press, 2015.

Worster, Donald. *Nature's Economy: A History of Ecological Ideas.* Second edition. Cambridge University Press, 1994. (First published in 1977.)

I. INTRATERRESTRIALS

Bomberg, Malin, and Lasse Ahonen. "Editorial: Geomicrobes: Life in Terrestrial Deep Subsurface." *Frontiers in Microbiology,* vol. 8, 2017, p. 103.

Borgonie, G., et al. "Nematoda from the Terrestrial Deep Subsurface of South Africa." *Nature,* vol. 474, 2011, pp. 79–82.

Borgonie, G., et al. "Eukaryotic Opportunists Dominate the Deep-Subsurface Biosphere in South Africa." *Nature Communications,* vol. 6, no. 8952, 2015.

Casar, Caitlin P. "Mineral-Hosted Biofilm Communities in the Continental Deep Subsurface, Deep Mine Microbial Observatory, SD, USA." *Geobiology,* vol. 18, no. 4, 2020, pp. 508–22.

Chivian, Dylan. "Environmental Genomics Reveals a Single-Species Ecosystem Deep Within Earth." *Science,* vol. 322, no. 5899, 2008, pp. 275–78.

Colman, Daniel R., et al. "The Deep, Hot Biosphere: A Retrospection." *Proceedings of the National Academy of Sciences,* vol. 114, no. 27, 2017, pp. 6895–6903.

Colwell, Frederick S., and Steven D'Hondt. "Nature and Extent of the Deep Biosphere." *Reviews in Mineralogy and Geochemistry,* vol. 75, no. 1, 2013, pp. 546–74.

Deep Carbon Observatory. "Deep Carbon Observatory: A Decade of Discovery." Deep Carbon Observatory Secretariat, Washington, D.C., 2019.

Eagle, Sina Bear. "The Lakota Emergence Story." National Park Service, 2019.

Edwards, K. J., et al. "The Deep, Dark Energy Biosphere: Intraterrestrial Life on Earth." *Annual Review of Earth and Planetary Sciences,* vol. 40, no. 1, 2012, pp. 551–68.

Gadd, Geoffrey Michael. "Metals, Minerals, and Microbes: Geomicrobiology and Bioremediation." *Microbiology,* vol. 156, 2010, pp. 609–43.

Grantham, Bill. *Creation Myths and Legends of the Creek Indians.* University Press of Florida, 2002.

Grosch, Eugene G., and Robert M. Hazen. "Microbes, Mineral Evolution, and the Rise of Microcontinents—Origin and Coevolution of Life with Early Earth." *Astrobiology,* vol. 15, no. 10, 2015, pp. 922–39.

Hazen, Robert M., editor. "Mineral Evolution." *Elements,* vol. 6, no. 1, 2010.

Hazen, Robert M. *Symphony in C: Carbon and the Evolution of (Almost) Everything.* W. W. Norton, 2019.

Hazen, Robert M., et al. "Mineral Evolution." *American Mineralogist,* vol. 93, 2008, pp. 1693–1720.

Holland, G., et al. "Deep Fracture Fluids Isolated in the Crust Since the Precambrian Era." *Nature,* vol. 497, 2013, pp. 357–60.

Höning, Dennis, et al. "Biotic vs. Abiotic Earth: A Model for Mantle Hydration and Continental Coverage." *Planetary and Space Science,* vol. 98, 2014, pp. 5–13.

Hunt, Will. *Underground: A Human History of the Worlds Beneath Our Feet.* Spiegel and Grau, 2019.

Lollar, Garnet S., et al. " 'Follow the Water': Hydrogeochemical Constraints on Microbial Investigations 2.4 km Below Surface at the Kidd Creek Deep Fluid and Deep Life Observatory." *Geomicrobiology Journal,* vol. 36, no. 10, 2019, pp. 859–72.

Mader, Brigitta. "Archduke Ludwig Salvator and Leptodirus Hohenwarti from Postojnska Jama." *Acta Carsologica,* vol. 32, no. 2, 2016.

Onstott, Tullis C. *Deep Life: The Hunt for the Hidden Biology of Earth, Mars, and Beyond.* Princeton University Press, 2017.

Osburn, Magdalena R., et al. "Establishment of the Deep Mine Microbial Observatory (DeMMO), South Dakota, USA, a Geochemically Stable Portal into the Deep Subsurface." *Frontiers in Earth Science,* vol. 7, no. 196, 2019.

Polak, Slavko. "Importance of Discovery of the First Cave Beetle: Leptodirus hochenwartii Schmidt, 1832." *Endins: publicació d'espeleologia* 28, 2005, pp. 71–80.

Rosing, Minik T., et al. "The Rise of Continents—An Essay on the Geologic Consequences of Photosynthesis." *Palaeogeography, Palaeoclimatology, Palaeoecology,* vol. 232, 2006, pp. 99–113.

Soares, A., et al. "A Global Perspective on Microbial Diversity in the Terrestrial Deep Subsurface." *bioRxiv,* 2019.

Southam, G., and James A. Saunders. "The Geomicrobiology of Ore Deposits." *Economic Geology,* vol. 100, no. 6, 2005, pp. 1067–84.

2. THE MAMMOTH STEPPE AND THE ELEPHANT'S FOOTPRINT

The account of the voyage to Wrangel Island is based on interviews with Nikita and Sergey Zimov and on Nikita's journal from the trip.

Anderson, Ross. "Welcome to Pleistocene Park." *The Atlantic,* April 2017.

Animal People, Inc. "An Interview with Nikita Zimov, Director of Pleistocene Park." *Animal People Forum,* April 2, 2017.

Bar-On, Yinon M., et al. "The Biomass Distribution on Earth." *Proceedings of the National Academy of Sciences,* vol. 115, no. 25, 2018, pp. 6506–11.

Bottjer, David J., et al. "The Cambrian Substrate Revolution." *GSA Today,* vol. 10, no. 9, 2000, pp. 1–7.

Buatois, L. A., et al. "Sediment Disturbance by Ediacaran Bulldozers and the Roots of the Cambrian Explosion." *Scientific Reports,* vol. 8, no. 4514, 2018.

Croft, B., et al. "Contribution of Arctic Seabird-Colony Ammonia to Atmospheric Particles and Cloud-Albedo Radiative Effect." *Nature Communications,* vol. 7, no. 13444, 2016.

Doughty, Christopher E., et al. "Biophysical Feedbacks Between the Pleistocene Megafauna Extinction and Climate: The First Human-Induced Global Warming?" *Geophysical Research Letters,* vol. 37, 2010.

Doughty, Christopher E., et al. "Global Nutrient Transport in a World of Giants." *Proceedings of the National Academy of Sciences,* vol. 113, no. 4, 2016, pp. 868–73.

Holdo, R. M., et al. "A Disease-Mediated Trophic Cascade in the Serengeti and Its Implications for Ecosystem C." *PLOS Biology,* vol. 7, no. 9, 2009.

Katija, Katani. "Biogenic Inputs to Ocean Mixing." *Journal of Experimental Biology,* vol. 215, 2012, pp. 1040–49.

Kintisch, Eli. "Born to Rewild." *Science,* December 2015.

Macias-Fauria M, et al. "Pleistocene Arctic Megafaunal Ecological Engineering as a Natural Climate Solution?" *Philosophical Transactions of the Royal Society B,* vol. 375, no. 1794, 2020.

Meysman, F. J., et al. "Bioturbation: A Fresh Look at Darwin's Last Idea." *Trends in Ecology and Evolution,* vol. 21, no. 12, 2006, pp. 688–95.

Payne, Jonathan L., et al. "The Evolution of Complex Life and the Stabilization of the Earth System." *Interface Focus,* vol. 10, no. 4, 2020.

Remmers, W., et al. "Elephant (*Loxodonta africana*) Footprints as Habitat for Aquatic Macroinvertebrate Communities in Kibale National Park, South-West Uganda." *African Journal of Ecology,* vol. 55, 2017, pp. 342–51.

Roman, Joe, and James J. McCarthy. "The Whale Pump: Marine Mammals Enhance Primary Productivity in a Coastal Basin." *PLOS ONE,* vol. 5, no. 10, 2010.

Schmitz, Oswald J., et al. "Animals and the Zoogeochemistry of the Carbon Cycle." *Science,* vol. 362, no. 6419, 2018.

Shapiro, Beth. *How to Clone a Mammoth.* Princeton University Press, 2015.

Shapiro, Beth, et al. "Rise and Fall of the Beringian Steppe Bison." *Science,* vol. 306, no. 5701, 2004, pp. 1561–65.

Vernadsky, Valdimir I. *The Biosphere: Complete Annotated Edition.* Copernicus, 1998.

Willis, K. J., and J. C. McElwain. *The Evolution of Plants.* Oxford University Press, 2014.

Wolf, Adam. "The Big Thaw." *Stanford,* September/October 2008.

Zimov, Nikita, et al. "Pleistocene Park: The Restoration of Steppes as a Tool to Mitigate Climate Change Through Albedo Effect." AGU Fall Meeting, 2017.

Zimov, Nikita, et al. "Pleistocene Park Experiment: Effect of Grazing on the Accumulation of Soil Carbon in the Arctic." AGU Fall Meeting, 2018.

Zimov, Sergey. "Mammoth Steppes and Future Climate." *Human Environment,* 2007.

Zimov, Sergey. *Wild Field Manifesto.* November 2014.

Zimov, Sergey, et al. "Steppe-Tundra Transition: A Herbivore-Driven Biome Shift at the End of the Pleistocene." *The American Naturalist,* vol. 146, no. 5, 1995, pp. 765–94.

Zimov, Sergey, et al. "The Past and Future of the Mammoth Steppe Ecosystem." *Paleontology in Ecology and Conservation,* edited by Julien Louys, pp. 193–225. Springer Earth System Sciences, 2012.

3. A GARDEN IN THE VOID

Angourakis, Andreas, et al. "Human-Plant Coevolution: A Modelling Framework for Theory-Building on the Origins of Agriculture." *PLOS ONE,* vol. 17, no. 9, 2022.

Arneth, Almut, et al. "Summary for Policymakers." *Climate Change and Land,* edited by P. R. Shukla et al. Intergovernmental Panel on Climate Change, 2019.

Borrelli, Pasquale, et al. "Land Use and Climate Change Impacts on Global Soil Erosion by Water (2015–2070)." *Proceedings of the National Academy of Sciences,* vol. 117, 2020, pp. 1–8.

Bradford, Mark, et al. "Soil Carbon Science for Policy and Practice." *Nature Sustainability,* vol. 2, no. 12, 2019, pp. 1070–72.

Broushaki, Farnaz, et al. "Early Neolithic Genomes from the Eastern Fertile Crescent." *Science,* vol. 353, no. 6298, 2016, pp. 499–503.

Chen, Le, et al. "The Impact of No-Till on Agricultural Land Values in the United States Midwest." *American Journal of Agricultural Economics,* vol. 105, no. 3, 2023, pp. 760–83.

Cotillon, Suzanne, et al. "Land Use Change and Climate-Smart Agriculture in the Sahel." *The Oxford Handbook of the African Sahel,* edited by Leonardo A. Villalón, pp. 209–30. Oxford Academic, 2021.

Dynarski, Katherine A., et al. "Dynamic Stability of Soil Carbon: Reassessing the

'Permanence' of Soil Carbon Sequestration." *Frontiers in Environmental Science,* vol. 8, 2020.

Eekhout, Joris P. C., and Joris de Vente. "Global Impact of Climate Change on Soil Erosion and Potential for Adaptation Through Soil Conservation." *Earth-Science Reviews,* vol. 226, 2022.

Erisman, Jan Willem, et al. "How a Century of Ammonia Synthesis Changed the World." *Nature Geoscience,* vol. 1, 2008, pp. 636–39.

Franzmeier, Donald P., et al. *Soil Science Simplified: Fifth Edition*. Waveland Press, 2016.

Giller, K. E., et al. "Regenerative Agriculture: An Agronomic Perspective." *Outlook on Agriculture,* vol. 50, no. 1, 2021, pp. 13–25.

Handelsman, Jo. *A World Without Soil: The Past, Present, and Precarious Future of the Earth Beneath Our Feet*. Yale University Press, 2021.

Hudson, Berman D. *Our Good Earth: A Natural History of Soil*. Algora Publishing, 2020.

Kassam, Amir, et al. "Successful Experiences and Lessons from Conservation Agriculture Worldwide." *Agronomy,* vol. 12, no. 4, 2022, p. 769.

Lal, Rattan, et al. "Evolution of the Plow over 10,000 Years and the Rationale for No-Till Farming." *Soil and Tillage Research,* vol. 93, 2007, pp. 1–12.

Lal, Rattan, et al. "The Carbon Sequestration Potential of Terrestrial Ecosystems." *Journal of Soil and Water Conservation,* vol. 73, no. 6, 2018, pp. 145A–152A.

Lehmann, Johannes, and Markus Kleber. "The Contentious Nature of Soil Organic Matter." *Nature,* vol. 528, 2015, pp. 60–68.

Levis, C., et al. "Persistent Effects of Pre-Columbian Plant Domestication on Amazonian Forest Composition." *Science,* vol. 355, 2017, pp. 925–31.

Marris, Emma. "A Call for Governments to Save Soil." *Nature,* vol. 601, 2022, pp. 503–4.

Montgomery, David. *Dirt: The Erosion of Civilizations*. University of California Press, 2007.

Montgomery, David. *Growing a Revolution: Bringing Our Soil Back to Life*. W. W. Norton, 2017.

Our World in Data. www.ourworldindata.org. Accessed 2023.

Pasiecznik, Nick, and Chris Reij, editors. *Restoring African Drylands*. Tropenbos International, 2020.

Paul, Eldor A. "The Nature and Dynamics of Soil Organic Matter: Plant Inputs, Microbial Transformations, and Organic Matter Stabilization." *Soil Biology and Biochemistry,* vol. 98, 2016, pp. 109–26.

Piccolo, Alessandro, et al. "The Molecular Composition of Humus Carbon: Recalcitrance and Reactivity in Soils." *The Future of Soil Carbon: Its Conservation and Formation,* edited by Carlos Garcia, Paolo Nannipieri, and Teresa Hernandez, pp. 87–124. Elsevier Academic Press, 2018.

Pingali, Prabhu. "Green Revolution: Impacts, Limits, and the Path Ahead." *Proceedings of the National Academy of Sciences,* vol. 109, 2012, pp. 12302–8.

Pollan, Michael. *Second Nature: A Gardener's Education*. Delta, 1991.

Retallack, Gregory J. *Soil Grown Tall: The Epic Saga of Life from Earth*. Springer, 2022.

Retallack, Gregory J., and Nora Noffke. "Are There Ancient Soils in the 3.7 Ga Isua

Greenstone Belt, Greenland?" *Palaeogeography, Palaeoclimatology, Palaeoecology,* vol. 514, 2019, pp. 18–30.

Roberts, Patrick, et al. "The Deep Human Prehistory of Global Tropical Forests and Its Relevance for Modern Conservation." *Nature Plants,* vol. 3, no. 8, 2017.

Sanderman, Jonathan, et al. "Soil Carbon Debt of 12,000 Years of Human Land Use." *Proceedings of the National Academy of Sciences,* vol. 114, no. 36, 2017, pp. 9575–80.

Schlesinger, William H., and Ronald Amundson. "Managing for Soil Carbon Sequestration: Let's Get Realistic." *Global Change Biology,* vol. 25, 2019, pp. 386–89.

Smil, Vaclac. *Enriching the Earth: Fritz Haber, Carl Bosch, and the Transformation of World Food Production*. The MIT Press, 2004.

Snir, Ainit, et al. "The Origin of Cultivation and Proto-Weeds, Long Before Neolithic Farming." *PLOS ONE,* vol. 10, no. 7, 2015.

Thaler, Evan A., et al. "The Extent of Soil Loss Across the US Corn Belt." *Proceedings of the National Academies of Sciences,* vol. 118, no. 8, 2021.

Weil, Ray R., and Nyle C. Brady. *The Nature and Properties of Soils: Fifteenth Edition*. Pearson, 2017.

Winkler, Karina, et al. "Global Land Use Changes Are Four Times Greater Than Previously Estimated." *Nature Communications,* vol. 12, no. 2501, 2021.

Zeder, Melinda A. "The Origins of Agriculture in the Near East." *Current Anthropology,* vol. 52, no. S4, 2011.

4. SEA CELLS

Ayers, Greg P., and Jill M. Cainey. "The CLAW Hypothesis: A Review of Major Developments." *Environmental Chemistry,* vol. 4, 2007, pp. 366–74.

Beaufort, Luc, et al. "Cyclic Evolution of Phytoplankton Forced by Changes in Tropical Seasonality." *Nature,* vol. 601, 2022, pp. 79–84.

Castellani, Claudia, and Martin Edwards, editors. *Marine Plankton: A Practical Guide to Ecology, Methodology, and Taxonomy*. Oxford University Press, 2017.

Chimileski, Scott, and Roberto Kolter. *Life at the Edge of Sight: A Photographic Exploration of the Microbial World*. Belknap Press, 2017.

De Wever, Patrick. *Marvelous Microfossils: Creators, Timekeepers, Architects*. John Hopkins University Press, 2020.

Deutsch, Curtis, and Thomas Weber. "Nutrient Ratios as a Tracer and Driver of Ocean Biogeochemistry." *Annual Review of Marine Science,* vol. 4, 2012, pp. 113–41.

Eichenseer, K., et al. "Jurassic Shift from Abiotic to Biotic Control on Marine Ecological Success." *Nature Geoscience,* vol. 12, 2019, pp. 638–42.

Falkowski, P. "Ocean Science: The Power of Plankton." *Nature,* vol. 483, 2012, pp. S17—S20.

Falkowski, Paul, and Andy Knoll, editors. *Evolution of Primary Producers in the Sea*. Elsevier Academic Press, 2007.

Green, Tamara K., and Angela D. Hatton. "The CLAW Hypothesis: A New Perspective on the Role of Biogenic Sulphur in the Regulation of Global Climate." *Oceanography and Marine Biology: An Annual Review,* vol. 52, no. 326, 2014, pp. 315–36.

Gruber, Nicolas. "The Dynamics of the Marine Nitrogen Cycle and its Influence on Atmospheric CO2 Variations." *The Ocean Carbon Cycle and Climate*. NATO Science Series (Series IV: Earth and Environmental Sciences), vol. 40, edited by M. Follows and T. Oguz, pp. 97–148. Springer, 2004.

Kirby, Richard R. *Ocean Drifters: A Secret World Beneath the Waves*. StudioCactus, 2010.

Nadis, Steve. "The Cells That Rule the Seas." *Scientific American*, December 2003.

Proctor, Robert. "A World of Things in Emergence and Growth: René Binet's Porte Monumentale at the 1900 Paris Exposition." *Symbolist Objects: Materiality and Subjectivity at the Fin-de-Siècle*, edited by Claire I. R. O'Mahony, pp. 224–49. Rivendale Press, 2009.

Ridgwell, Andy, and Richard E. Zeebe. "The Role of the Global Carbonate Cycle in the Regulation and Evolution of the Earth System." *Earth and Planetary Science Letters:* vol. 234, no. 3–4, 2005, pp. 299–315.

Rohling, Eelco J. *The Oceans: A Deep History*. Princeton University Press, 2017.

Sardet, Christian. *Plankton: Wonders of the Drifting World*. University of Chicago Press, 2015.

Yu, Hongbin, et al. "The Fertilizing Role of African Dust in the Amazon Rainforest: a First Multiyear Assessment Based on Data from Cloud Aerosol Lidar and Infrared Pathfinder Satellite Observations." *Geophysical Research Letters*, vol. 42, 2015, pp. 1984–91.

5. THESE GREAT AQUATIC FORESTS

Chapman, R. L. "Algae: The World's Most Important 'Plants'—An Introduction." *Mitigation and Adaptation Strategies for Global Change*, vol. 18, 2013, pp. 5–12.

Delaney, A., et al. "Society and Seaweed: Understanding the Past and Present." *Seaweed in Health and Disease Prevention*, edited by Joël Fleurence and Ira Levine, pp. 7–40. Elsevier Academic Press, 2016.

Dillehay, Tom D., et al. "Monte Verde: Seaweed, Food, Medicine, and the Peopling of South America." *Science*, vol. 320, no. 5877, 2008, pp. 784–86.

Duarte, Carlos. "Reviews and Syntheses: Hidden Forests, the Role of Vegetated Coastal Habitats in the Ocean Carbon Budget." *Biogeosciences*, vol. 14, no. 2, 2017, pp. 301–10.

Duarte, Carlos, et al. "Can Seaweed Farming Play a Role in Climate Change Mitigation and Adaptation?" *Frontiers in Marine Science*, vol. 4, 2017.

Eckman, James E., et al. "Ecology of Understory Kelp Environments. I. Effects of Kelps on Flow and Particle Transport near the Bottom." *Journal of Experimental Marine Biology and Ecology*, vol. 129, no. 2, 1989, pp. 173–87.

Flannery, Tim. *Sunlight and Seaweed: An Argument for How to Feed, Power, and Clean Up the World*. Text Publishing, 2017.

Hurd, Catriona L., et al. *Seaweed Ecology and Physiology: Second Edition*. Cambridge University Press, 2014.

Langton, Richard, et al. "An Ecosystem Approach to the Culture of Seaweed." Tech. Memo. NMFS-F/SPO-195, 24, National Oceanic and Atmospheric Administration, 2019.

Mouritsen, Ole. "The Science of Seaweeds." *American Scientist*, 2013.

Naar, Nicole. "Puget Sound Kelp Conservation and Recovery Plan: Appendix B—The Cultural Importance of Kelp for Pacific Northwest Tribes." National Oceanic and Atmospheric Administration, May 2020.

Nielsen, Karina J., et al. "Emerging Understanding of the Potential Role of Seagrass and Kelp as an Ocean Acidification Management Tool in California." California Ocean Science Trust, Oakland, California, January 2018.

Nisizawa, K., et al. "The Main Seaweed Foods in Japan." *Hydrobiologia,* vol. 151, 1987, pp. 5–29.

O'Connor, Kaori. *Seaweed: A Global History.* Reaktion Books, 2017.

Ortega, A., et al. "Important Contribution of Macroalgae to Oceanic Carbon Sequestration." *Nature Geoscience,* vol. 12, 2019, pp. 748–54.

Pfister, C. A., et al. "Kelp Beds and Their Local Effects on Seawater Chemistry, Productivity, and Microbial Communities." *Ecology,* vol. 100, no. 10, 2019.

Proceedings of the First U.S.-Japan Meeting on Aquaculture at Tokyo, Japan, October 18–19, 1971: Under the U.S.-Japan Cooperative Program in Natural Resources (UJNR). Edited by William N. Shaw. National Marine Fisheries Service, National Oceanic and Atmospheric Administration, U.S. Department of Commerce, 1974.

Puget Sound Restoration Fund. "Summary of Findings: Investigating Seaweed Cultivation as a Strategy for Mitigating Ocean Acidification in Hood Canal, WA." 2019.

Rosman, Johanna H., et al. "Currents and Turbulence Within a Kelp Forest (*Macrocystis Pyrifera*): Insights from a Dynamically Scaled Laboratory Model." *Limnology and Oceanography,* vol. 55, 2010, pp. 1145–58.

Shetterly, Susan Hand. *Seaweed Chronicles: A World at the Water's Edge.* Algonquin Books, 2018.

Tripati, Robert Eagle, et al. "Kelp Forests as a Refugium: A Chemical and Spatial Survey of a Palos Verdes Restoration Area: Project Report." UCLA Environmental Science Practicum, 2016–2017.

Wiencke, Christian, and Kai Bischof, editors. *Seaweed Biology: Novel Insights into Ecophysiology, Ecology, and Utilization.* Springer, 2012.

6. PLASTIC PLANET

Borunda, Alejandra. "This Young Whale Died with 88 Pounds of Plastic in Its Stomach." *National Geographic,* March 18, 2019.

Case, Emalani. "Caught (and Brought) in the Currents: Narratives of Convergence, Destruction, and Creation at Kamilo Beach." *Journal of Transnational American Studies,* vol. 10, no. 1, 2019, pp. 73–92.

Corcoran, Patricia L., et al. "An Anthropogenic Marker Horizon in the Future Rock Record." *GSA Today,* vol. 24, no. 6, 2014, pp. 4–8.

Cox, Kieran D., et al. "Human Consumption of Microplastics." *Environmental Science and Technology,* vol. 53, no. 12, 2019, pp. 7068–74.

De-la-Torre, Gabriel Enrique, et al. "New Plastic Formations in the Anthropocene." *Science of the Total Environment,* vol. 754, 2021.

Freinkel, Susan. "A Brief History of Plastic's Conquest of the World." *Scientific American,* May 29, 2011.

Gabbott, Sarah, et al. "The Geography and Geology of Plastics: Their Environmental Distribution and Fate." *Plastic Waste and Recycling: Environmental Impact, Societal Issues, Prevention, and Solutions,* edited by Trevor M. Letcher, pp. 33–63. Academic Press, 2020.

Geyer, Roland. "A Brief History of Plastics." *Mare Plasticum: The Plastic Sea,* edited by Marilena Streit-Bianchi et al., pp. 31–48. Springer, 2020.

Geyer, Roland, et al. "Production, Use, and Fate of All Plastics Ever Made." *Science Advances,* vol. 3, no. 7, 2017.

Hamilton, Lisa Anne, and Steven Feit et al. "Plastic and Climate: The Hidden Costs of a Plastic Planet." Center for International Environmental Law, 2019.

Haram, Linsey E. "Emergence of a Neopelagic Community Through the Establishment of Coastal Species on the High Seas." *Nature Communications,* vol. 12, no. 1, 2021.

Jenner, Lauren C., et al. "Detection of Microplastics in Human Lung Tissue Using μFTIR Spectroscopy." *Science of the Total Environment,* vol. 831, 2022.

Meijer, Lourens J. J., et al. "Over 1000 Rivers Accountable for 80% of Global Riverine Plastic Emissions into the Ocean." *Science Advances,* vol. 7, no. 18, 2021.

Moore, Charles. "Trashed: Across the Pacific Ocean, Plastics, Plastics Everywhere." *Natural History,* vol. 112, no. 9, 2003, pp. 46–51.

Moore, C. J., et al. "A Comparison of Plastic and Plankton in the North Pacific Central Gyre." *Marine Pollution Bulletin,* vol. 42, no. 12, 2001, pp. 1297–1300.

Moore, Charles, and Cassandra Philips. *Plastic Ocean: How a Sea Captain's Chance Discovery Launched a Determined Quest to Save the Oceans.* Avery, 2011.

Our World in Data. www.ourworldindata.org. Accessed 2022.

PEW Charitable Trusts and SystemIQ. "Breaking the Plastic Wave: A Comprehensive Assessment of Pathways Towards Stopping Ocean Plastic Pollution." 2020.

Raworth, Kate. *Doughnut Economics: Seven Ways to Think Like a 21st-Century Economist.* Chelsea Green Publishing, 2017.

Shen, Maocai, et al. "Can Microplastics Pose a Threat to Ocean Carbon Sequestration?" *Marine Pollution Bulletin,* vol. 150, 2020.

Tarkanian, Michael J., and Dorothy Hosler. "America's First Polymer Scientists: Rubber Processing, Use and Transport in Mesoamerica." *Latin American Antiquity,* vol. 22, no. 4, 2011, pp. 469–86.

Watt, Ethan. "Ocean Plastics: Environmental Implications and Potential Routes for Mitigation—A Perspective." *RSC Advances,* vol. 11, no. 35, 2021, pp. 21447–62.

Wayman, Chloe, and Helge Niemann. "The Fate of Plastic in the Ocean Environment—A Minireview." *Environmental Science: Processes and Impacts,* vol. 23, 2021, pp. 198–212.

Worm, Boris, et al. "Plastic as a Persistent Marine Pollutant." *Annual Review of Environment and Resources,* vol. 42, 2017, pp. 1–26.

Wright, Robyn J., et al. "Marine Plastic Debris: A New Surface for Microbial Colonization." *Environmental Science and Technology,* vol. 54, no. 19, 2020, pp. 11657–72.

Yoshida, Shosuke, et al. "A Bacterium That Degrades and Assimilates Poly(ethylene Terephthalate)." *Science,* vol. 351, no. 6278, 2016, pp. 1196–99.

Zalasiewicz, Jan, et al. "The Geological Cycle of Plastics and Their Use as a Stratigraphic Indicator of the Anthropocene." *Anthropocene,* vol. 13, 2016, pp. 4–17.

7. A BUBBLE OF BREATH

Alcott, Lewis J., et al. "Stepwise Earth Oxygenation Is an Inherent Property of Global Biogeochemical Cycling." *Science,* vol. 366, no. 6471, 2019, pp. 1333–37.

Andreae, Meinrat, et al. "The Amazon Tall Tower Observatory (ATTO): Overview of Pilot Measurements on Ecosystem Ecology, Meteorology, Trace Gases, and Aerosols." *Atmospheric Chemistry and Physics,* vol. 15, no. 18, 2015, pp. 10723–76.

DeLeon-Rodriguez, Natasha, et al. "Microbiome of the Upper Troposphere: Species Composition and Prevalence, Effects of Tropical Storms, and Atmospheric Implications." *Proceedings of the National Academy of Sciences,* vol. 110, no. 7, 2013, pp. 2575–80.

Fröhlich-Nowoisky, Janine, et al. "Bioaerosols in the Earth System: Climate, Health, and Ecosystem Interactions." *Atmospheric Research,* vol. 182, 2016, pp. 346–76.

Gumsley, Ashley, et al. "Timing and Tempo of the Great Oxidation Event." *Proceedings of the National Academy of Sciences,* vol. 114, no. 8, 2017, pp. 1811–16.

Krause, A. J., et al. "Stepwise Oxygenation of the Paleozoic Atmosphere." *Nature Communications,* vol. 9, no. 4081, 2018.

Lenton, Timothy, et al. "Co-evolution of Eukaryotes and Ocean Oxygenation in the Neoproterozoic Era." *Nature Geoscience,* vol. 7, 2014, pp. 257–65.

Lovejoy, Thomas E., and Carlos Nobre. "Amazon Tipping Point." *Science Advances,* vol. 4, no. 2, 2018.

Lovejoy, Thomas E., and Carlos Nobre. "Amazon Tipping Point: Last Chance for Action." *Science Advances,* vol. 5, no. 12, 2019.

Lyons, Timothy, et al. "Oxygenation, Life, and the Planetary System During Earth's Middle History: An Overview." *Astrobiology,* vol. 21, no. 8, 2021, pp. 906–23.

Mills, Daniel, et al. "Eukaryogenesis and Oxygen in Earth History." *Nature Ecology and Evolution,* vol. 6, 2022, pp. 520–32.

Morris, Cindy E., et al. "Bioprecipitation: A Feedback Cycle Linking Earth History, Ecosystem Dynamics, and Land Use Through Biological Ice Nucleators in the Atmosphere." *Global Change Biology,* vol. 20, no. 2, 2014, pp. 341–51.

Olejarz, Jason, et al. "The Great Oxygenation Event as a Consequence of Ecological Dynamics Modulated by Planetary Change." *Nature Communications,* vol. 12, no. 3985, 2021.

Ostrander, Chadlin M., et al. "Earth's First Redox Revolution." *Annual Review of Earth and Planetary Sciences,* vol. 49, 2021, pp. 337–66.

Pöhlker, Christopher, et al. "Biogenic Potassium Salt Particles as Seeds for Secondary Organic Aerosol in the Amazon." *Science,* vol. 337, 2012, pp. 1075–78.

Pöschl, Ulrich, et al. "Rainforest Aerosols as Biogenic Nuclei of Clouds and Precipitation in the Amazon." *Science,* vol. 329, no. 5998, 2010, pp. 1513–16.

Sánchez-Baracaldo, Patricia, et al. "Cyanobacteria and Biogeochemical Cycles Through Earth History." *Trends in Microbiology,* vol. 30, no. 2, 2022, pp. 143–57.

Soubeyrand, Samuel, et al. "Analysis of Fragmented Time Directionality in Time Series to Elucidate Feedbacks in Climate Data." *Environmental Modelling and Software,* vol. 61, 2014, pp. 78–86.

Sperling, Erik A., et al. "Oxygen, Ecology, and the Cambrian Radiation of Animals." *Proceedings of the National Academy of Sciences of the United States of America,* vol. 110, no. 33, 2013, pp. 13446–51.

Steffen, Will, et al. "The Emergence and Evolution of Earth System Science." *Nature Reviews Earth and Environment,* vol. 1, 2020, pp. 54–63.

Upper, Christen D., and Gabor Vali. "Chapter 2: The Discovery of Bacterial Ice Nucleation and the Role of Bacterial Ice Nucleation in Frost Injury to Plants." *Biological Ice Nucleation and Its Applications,* edited by R. E. Lee, Jr., and G. J. Warren, pp. 29–40. APS Press, 1995.

8. THE ROOTS OF FIRE

Alcott, Lewis J., et al. "Stepwise Earth Oxygenation Is an Inherent Property of Global Biogeochemical Cycling." *Science,* vol. 366, no. 6471, 2019, pp. 1333–37.

Anderson, M. Kat. "The Use of Fire by Native Americans in California." *Fire in California's Ecosystems,* edited by Neil G. Sugihara et al., pp. 417–30. University of California Press, 2006.

Beerling, David. *The Emerald Planet: How Plants Changed Earth's History.* Oxford University Press, 2007.

Bouchard, F. "Ecosystem Evolution is About Variation and Persistence, not Populations and Reproduction." *Biological Theory,* vol. 9, 2014, pp. 382–91.

Bowman, David M.J.S., et al. "Fire in the Earth System." *Science,* vol. 324, no. 5926, 2009, pp. 481–84.

David, A. T., Asarian, J. E., and Lake, F. K. "Wildfire Smoke Cools Summer River and Stream Water Temperatures." *Water Resources Research,* vol. 54, 2018, pp. 7273–90.

Doolittle, W. Ford. "Is the Earth an Organism?" *Aeon,* December 2020.

Doolittle, W. Ford, and S. Andrew Inkpen. "Processes and Patterns of Interaction as Units of Selection: An Introduction to ITSNTS Thinking." *Proceedings of the National Academy of Sciences,* vol. 115, no. 16, 2018, pp. 4006–14.

Dussault, Antoine C., and Frédéric Bouchard. "A Persistence Enhancing Propensity Account of Ecological Function to Explain Ecosystem Evolution." *Synthese,* vol. 194, 2017, pp. 1115–45.

Hazen, Robert M. *Symphony in C: Carbon and the Evolution of (Almost) Everything.* W. W. Norton, 2019.

Judson, Olivia. "The Energy Expansions of Evolution." *Nature Ecology and Evolution,* vol. 1, no. 138, 2017.

Kay, Charles E. "Native Burning in Western North America: Implications for Hardwood Forest Management." *Proceedings: Workshop on Fire, People, and the Central Hardwoods Landscape,* compiled by Daniel A. Yaussy, Richmond, Kentucky, March 12–14, 2000.

Krause, A. J., et al. "Stepwise Oxygenation of the Paleozoic Atmosphere." *Nature Communications,* vol. 9, no. 4081, 2018.

Kump, L. R. "Terrestrial Feedback in Atmospheric Oxygen Regulation by Fire and Phosphorus." *Nature,* vol. 335, 1988, pp. 152–54.

Lake, Frank K. *Traditional Ecological Knowledge to Develop and Maintain Fire Regimes in Northwestern California, Klamath-Siskiyou Bioregion: Management and Restoration of Culturally Significant Habitats.* PhD dissertation, Oregon State University, 2007.

Lenton, Timothy M. "The Role of Land Plants, Phosphorus Weathering, and Fire in the Rise and Regulation of Atmospheric Oxygen." *Global Change Biology,* vol. 7, 2001, pp. 613–29.

Lenton, Timothy M., et al. "First Plants Oxygenated the Atmosphere and Ocean." *Proceedings of the National Academy of Sciences,* vol. 113, no. 35, 2016, pp. 9704–9.

Lenton, Timothy M., et al. "Life on Earth Is Hard to Spot." *The Anthropocene Review,* vol. 7, no. 3, 2020, pp. 248–72.

Lenton, Timothy M., et al. "Survival of the Systems." *Trends in Ecology and Evolution,* vol. 36, no. 4, 2021, pp. 333–44.

McGhee, Jr., George R. *Carboniferous Giants and Mass Extinction: The Late Paleozoic Ice Age World.* Columbia University Press, 2018.

Pyne, Stephen J. *Fire: A Brief History.* University of Washington Press, 2001.

Pyne, Stephen J. "The Ecology of Fire." *Nature Education Knowledge,* vol. 3, no. 10, 2010.

Stanley, Steven M., and John A. Luczaj. *Earth System History.* Fourth edition. W. H. Freeman, 2015.

Williams, Gerald W. "References on the American Indian Use of Fire in Ecosystems." United States Forest Service, United States Department of Agriculture, 2005.

Willis, K. J., and J. C. McElwain. *The Evolution of Plants.* Oxford University Press, 2014.

9. WINDS OF CHANGE

Archer, David. *The Long Thaw: How Humans Are Changing the Next 100,000 Years of Earth's Climate.* Princeton University Press, 2009.

Cuddington, Kim. "The 'Balance of Nature' Metaphor and Equilibrium in Population Ecology." *Biology and Philosophy,* vol. 16, 2001, pp. 463–79.

Dessler, Andrew. *Introduction to Modern Climate Change.* Second edition. Cambridge University Press, 2016.

Egerton, Frank N. "Changing Concepts of the Balance of Nature." *Quarterly Review of Biology,* vol. 48, no. 2, 1973, pp. 322–50.

Freese, Barbara. *Coal: A Human History.* Perseus Publishing, 2003.

Jelinski, Dennis. "There Is No Mother Nature—There Is No Balance of Nature: Culture, Ecology, and Conservation." *Human Ecology,* vol. 33, no. 2, 2005, pp. 271–88.

Maslin, Mark. *Global Warming: A Very Short Introduction.* Oxford University Press, 2009.

Maslin, Mark, et al. "New Views on an Old Forest: Assessing the Longevity, Resilience, and Future of the Amazon Rainforest." *Transactions of the Institute of British Geographers,* vol. 30, no. 4, 2005, pp. 477–99.

Otto, Friederike E. L., et al. "Climate Change Likely Increased Extreme Monsoon Rainfall, Flooding Highly Vulnerable Communities in Pakistan." *World Weather Attribution,* September 2022.

Our World in Data. www.ourworldindata.org. Accessed 2023.

Pörtner, H.-O., et al. "Climate Change 2022: Impacts, Adaptation, and Vulnerability. Contribution of Working Group II to the Sixth Assessment Report of the

Intergovernmental Panel on Climate Change." Cambridge University Press, in press.

Schobert, Harold H. *The Chemistry of Fossil Fuels and Biofuels*. Cambridge University Press, 2013.

Shukla, P. R., et al. *Climate Change 2022: Mitigation of Climate Change*. Contribution of Working Group III to the Sixth Assessment Report of the Intergovernmental Panel on Climate Change. Cambridge University Press, 2022.

Simberloff, Daniel. "The 'Balance of Nature'—Evolution of a Panchreston." *PLOS Biology*, vol. 12, no. 10, 2014.

Smil, Vaclav. *Oil: A Beginner's Guide*. Second edition. Oneworld Publications, 2008.

Smil, Vaclav. *Energy Transitions: Global and National Perspectives*. Second edition. Praeger, 2017.

Smil, Vaclav. *Grand Transitions: How the Modern World Was Made*. Oxford University Press, 2021.

BECOMING EARTH

FERRIS JABR

A Book Club Guide

FASCINATING FACTS

1. The majority of the planet's microbes—perhaps more than 90 percent—may live deep underground, within the planet's rocky crust.

2. In North America, bison help engineer spring. As they intensively graze and fertilize grass, they encourage the plants to continually produce palatable and nutritious young shoots, propelling waves of springtime rejuvenation across the plains. Collectively, bison exert a stronger influence on seasonal plant growth than weather or other environmental factors.

3. In South America, some species of leafcutter ant construct underground nests that span thousands of square feet and extend as much as twenty-six feet deep, requiring them to move more than forty tons of soil.

4. The oldest water ever found on Earth, extracted from deep underground, has the consistency of maple syrup, contains twice as much salt as seawater, and tastes terrible.

5. Our planet typically requires centuries to create a single inch of fertile topsoil. Most of Earth's soils formed over tens of thousands of years, many over hundreds of thousands of years, and some over millions.

6. Altogether, the planet's soils store somewhere between 2.5 and 3 trillion tons of carbon, which is around three times more than all the carbon in the atmosphere and about four times as much as in all living vegetation.

7. A tablespoon of healthy soil holds a population of organisms that is easily many times the number of humans alive today. A single gram of fertile soil may contain billions of microbes and viruses, millions of protozoans and algae, hundreds of nematodes, dozens of mites and springtails, and a thousand meters of filamentous fungi.

8. As of 2020, researchers have documented more than 430 species that can digest various forms of plastic. Most are bacteria or fungi, but there are also some insect larvae in this growing group of plastivores.

9. Plankton are primarily responsible for seafoam and the funky signature smell of the ocean. When plankton blooms dwindle and perish, the wind and waves often mix their decomposing proteins and fats with other bits of organic detritus, such as fragments of coral, seaweed, and fish scales. This moldering mélange acts as a foaming agent, generating numerous air bubbles that balloon into a thick froth, a kind of plankton meringue, which washes onto shore. Meanwhile, the sulfur aerosols generated by dying and decomposing plankton give sea air much of its characteristic funk, an odor reminiscent of boiled beets. That scent mingles with briny bromophenols, produced in large quantities by marine worms and algae, and the strong "ocean smell" of certain seaweed sex pheromones. On a sterile planet, the seaside would not smell like the sea—at least not as we know it. When you breathe in sea air, you are literally breathing in sea life.

10. Otters use kelp as a kind of toddler leash, wrapping their pups in seaweed to prevent them from drifting away while they track down a meal.

11. Tiny particles of sinking ocean debris, known as marine snow, are composed in large part of plankton and accumulate on about 60 percent of the seafloor today. The uppermost layers of these sediments are like slurries, almost fluffy in texture. A few feet down, as the pressure increases, squeezing out water, they develop the consistency of toothpaste. Eventually, they are compressed into rock and are either melted in Earth's interior or returned to the surface by, say, clashing continental plates or shrinking seas.

12. The White Cliffs of Dover are more than just rock—they are also fossils. The cliffs are primarily made of the compressed skeletal remains of single-celled plankton called coccolithophores that lived during the Cretaceous Period, between 145 and 66 million years ago. In fact, the vast majority of chalk and limestone formations on Earth, including large sections of the Alps, are the remains of plankton, corals, shellfish, and other calcareous sea creatures. Every imposing edifice that humans have constructed with limestone, including the Great Pyramid of Giza, the Colosseum, Notre-Dame cathedral, and the Empire State Building, is a secret monument to ancient ocean life.

13. Like coral reefs, mangroves, and marshes, kelp forests shield coastal communities from the brunt of storms, reducing wave heights by up to 60 percent.

14. In 2014, scientists formally proposed a name for a new type of rock: *plastiglomerate,* the first rock type in the history of Earth composed partially of plastic

15. Every year, the wind carries immense quantities of Saharan dust across the Atlantic Ocean, depositing 27.7 million tons—enough to fill more than a hundred thousand semi-trailer trucks—in the Amazon rainforest, where it provides trillions of plants with iron, phosphorus, and other essential nutrients. This fertilizing dust is not simply tiny bits of dirt and rock; it is largely composed of the skeletons of ancient single-celled ocean plankton.

16. The sky is blue because of life. Ancient Earth likely had a hazy orange atmosphere with essentially no oxygen. Cyanobacteria, land plants, and other photosynthetic life oxygenated the atmosphere, transforming its chemistry and the way it interacts with light, which in turn shifted the sky's hue toward the blue end of the visible spectrum.

17. The Amazon rainforest generates about half of the rain that falls on its canopy each year. Rainforests pull huge volumes of water from the soil and release what they do not use to the atmosphere. In parallel, they emit invisible plumes of tiny airborne particles on which that water can condense: a complex mixture of organisms and organic entities, including viruses, microbes, algae, and pollen grains; the spores of fungi, mosses, and ferns; bits of leaves and bark; flecks of fur and feather; and slivers of scales of insect shells. This levitating assemblage of life and its vestiges can seed both clouds and ice crystals within clouds, significantly increasing the likelihood of precipitation and the pace of the water cycle.

18. The Karuk teach that fire can "call back the salmon from the ocean," but many scientists dismissed this as folklore. Using NASA satellite imagery and meteorological records, however, Indigenous fire ecologist Frank Lake and two colleagues demonstrated that by reflecting heat and light, wildfire smoke lowers the temperature of rivers, improving the survival of migrating salmon and other cold water–adapted species, especially during heat waves.

19. A single gallon of gasoline represents *one hundred tons* of ancient life, roughly equal to twenty adult elephants. Every sedan with a typical fifteen-gallon gas tank demands the equivalent of three hundred elephants simply to keep running. Fossil fuels are not just conveniently concentrated forms of energy—they are outrageously extravagant. A fossil fuel is essentially an ecosystem in an urn.

QUESTIONS FOR DISCUSSION

1. What is the Gaia hypothesis (p. xvi)? Who conceived it? And how has the scientific community's attitude towards Gaia changed over time?
2. Does science have a consensus definition of life (p. xviii)? Why or why not? What does Jabr single out as life's most important characteristic? Do you agree?
3. In what way is a living planet an emergent phenomenon (p. xx)? Why does Jabr draw a comparison between a living planet and music? What are other examples of complex phenomena that emerge from interactions between their constituent elements?
4. How does the Amazon make its own rain (p. xii)? How has our understanding of this process advanced in recent decades? And how does it challenge the typical way we think about the relationship between life and the planet?
5. What are particularly astonishing examples of animals that engineer their ecosystems (p. 38)? How does the Cambrian substrate revolution demonstrate both the importance and the longevity of such changes (p. 41)? When ancient humans hunted numerous megafauna species to extinction, how did they disrupt Earth's ecology at large (p. 28)? Are similar ecological disturbances occurring today?

6. The microbes that inhabit the planet's crust also change it. What are some of the most important ways in which they do so (pp. 17–20)? How are microorganisms involved in the processes that form limestone caves (p. 17)? How did microbes and other life forms create the necessary conditions for more than half of Earth's five thousand unique mineral species (p. 18)? Some scientists have even proposed that microbes played an important role in the formation of the continents (p. 19). What is the basis of this idea? How might microscopic creatures affect such a major geological process? How might Earth have developed differently if ancient microbial life had not existed?

7. In the opening of chapter 3, Jabr describes how his relationship with gardening has evolved throughout his life and the major challenges he encountered when trying to create an entire garden from the ground up (pp. 49–52). What have been your experiences with gardening? If you currently have a yard or garden, how would you characterize your relationship with it?

8. How much of the planet's habitable land have humans commandeered to grow crops or raise livestock (p. 52)? What are the major ecological consequences of this massive change to the planet's land surfaces? What ancient technology does Jabr highlight as both extremely important for the development of agriculture and ultimately detrimental to global ecology? What is the Haber-Bosch process and why is it such a milestone in both human history and the history of life (p. 55)? How have industrial agriculture and synthetic fertilizers warped the nitrogen cycle?

9. In the concluding section to chapter 5, Jabr writes that as our species struggles to manage the climate crisis, we repeatedly "find that we know enough to recognize and even quantify the importance of the astoundingly complex ecosystems we inhabit but not always enough to confidently intervene when they begin to collapse" (p. 119). How do seaweed-based carbon-capture schemes illustrate this tension? What are some other examples? What does

Jabr highlight as reasons for hope and perseverance despite these challenges? Given what you have learned about the Earth system so far, what gives you courage, solace, or inspiration?

10. Jabr writes that "turn off the tap" is a favorite mantra among scientists and environmentalists when discussing solutions to the plastic pollution crisis (p. 139). What does this phrase mean? How are conservationists and entrepreneurs attempting to curb, contain, or collect plastic waste? Which of these strategies are most and least effective, and why (p. 140)? Can you think of alternative approaches? What changes can you make in your own life to limit your use of disposable plastic?

11. In the conclusion to chapter 7, Jabr writes that most people "would not hesitate" to describe a forest as alive (p. 169). Do you agree with this characterization? In what way is a forest or other ecosystem alive and how is it different from a cell or organism? Why does Jabr underscore that Earth is not an organism or superorganism but rather a vast living system? How is this perspective more accurate and clarifying than those that have previously been associated with the Gaia hypothesis? What is the basis for Jabr's claim that life does not simply inhabit the planet but is in fact materially continuous with the planet?

12. When European colonists arrived in western North America, "they often encountered beautiful, parklike mosaics of forest and grassland, so open and spacious that they could easily maneuver horse-drawn carriages through them" (p. 172). They mistook these landscapes for pristine wilderness. What was their true origin? How did Indigenous Peoples deliberately use fire to shape their environments? What were some of the main benefits of doing so (p. 173)? What happened to North America's forests and other ecosystems when such practices were suppressed and outlawed? Why is it so important to bring them back (p. 176)?

13. How exactly do fossil fuels and other sources of greenhouse gases cause global warming (p. 195)? What are the major consequences

of this rise in global average temperature (p. 201)? What happens to the atmosphere's ability to hold water as the planet warms? Why is sea level rising? What are some of the repercussions for nonhuman species? What about consequences for agriculture and public health? Can we rely on Earth's innate self-regulating processes, such as the planetary thermostat, to restabilize global climate on their own (p. 203)? Why or why not?

14. Why has Western science been so dismissive of the concept of balance in nature (p. 215)? Does this disdain mask important truths about our world, and if so how? Given everything you have learned about the Earth system, do you think it's accurate to describe the planet as tending toward balance? How is it possible for balance to be, as Rachel Carson phrased it, "fluid, ever shifting, in a constant state of adjustment" (p. 217)? What are some specific examples of this kind of dynamic balance at the levels of organisms, ecosystems, and planets?

15. In the epilogue, Jabr identifies "three opposing perspectives on the fate of humanity and our planet": the fatalists, the fantasists, and the futurists (p. 227). What distinguishes the three from each other? What fourth alternative perspective does Jabr offer and how does he support his argument? Which one of these viewpoints do you agree with most? Or do you prefer another outlook altogether?

16. "Earth has shown us the power of community, diversity, and reciprocity," Jabr writes in the epilogue (p. 229). "Among all existing creatures, we alone have the opportunity to consciously emulate our living planet and knowingly perpetuate its sublime composition. We are neither the cancer of Earth nor its cure. We are its progeny, its poetry, and its mirror." How do you interpret this passage? Do you agree? What is the significance of the final passage, in which Jabr exchanges breath with the planet? How has this book changed the way you think about the relationship between Earth and life? Do you see the world differently now?

INDEX

Ferris Jabr is a contributing writer for *The New York Times Magazine* and *Scientific American*. He has also written for *The New Yorker, Harper's, The Atlantic, National Geographic, Foreign Policy, Wired, Outside, Lapham's Quarterly, McSweeney's,* and the *Los Angeles Review of Books,* among other publications. He is a recipient of a Whiting Foundation Creative Nonfiction Grant and fellowships from UC Berkeley and the MIT Knight Science Journalism Program. His work has been anthologized in the 2014, 2020, and 2023 editions of *Best American Science and Nature Writing*. He lives in Portland, Oregon, with his husband, Ryan, their dog, Jack, and more plants than they can count.

ABOUT THE TYPE

This book was set in Bembo, a typeface based on an old-style Roman face that was used for Cardinal Pietro Bembo's tract *De Aetna* in 1495. Bembo was cut by Francesco Griffo (1450–1518) in the early sixteenth century for Italian Renaissance printer and publisher Aldus Manutius (1449–1515). The Lanston Monotype Company of Philadelphia brought the well-proportioned letterforms of Bembo to the United States in the 1930s.